New Directions in Philosophy and Cognitive Science

Series Editor
Michelle Maiese
Admin 451
Emmanuel College
Boston, MA, USA

This series brings together work that takes cognitive science in new directions. For many years, philosophical contributions to the field of cognitive science came primarily from theorists with commitments to physical reductionism, neurocentrism, and a representationalist model of the mind. However, over the last two decades, a rich literature that challenges these traditional views has emerged. According to so-called '4E' approaches, the mind is embodied, embedded, enactive, and extended. Cognition, emotion, and consciousness are not best understood as comprised of brainbound representational mechanisms, but rather as dynamic, embodied, action-oriented processes that sometimes extend beyond the human body. Such work often draws from phenomenology and dynamic systems theory to rethink the nature of cognition, characterizing it in terms of the embodied activity of an affectively attuned organism embedded in its social world. In recent years, theorists have begun to utilize 4E approaches to investigate questions in philosophy of psychiatry, moral psychology, ethics, and political philosophy. To foster this growing interest in rethinking traditional philosophical notions of cognition using phenomenology, dynamic systems theory, and 4E approaches, we dedicate this series to "New Directions in Philosophy and Cognitive Science."

If you are interested in the series or wish to submit a proposal, please contact Amy Invernizzi, amy.invernizzi@palgrave-usa.com.

More information about this series at
http://link.springer.com/series/14744

Victor Loughlin

4E Cognitive Science and Wittgenstein

palgrave
macmillan

Victor Loughlin
Department of Philosophy
University of Antwerp
Antwerp, Belgium

New Directions in Philosophy and Cognitive Science
ISBN 978-3-030-89462-7   ISBN 978-3-030-89463-4 (eBook)
https://doi.org/10.1007/978-3-030-89463-4

This Palgrave Macmillan imprint is published by the registered company Springer Nature Switzerland AG.
The registered company address is: Gewerbestrasse 11, 6330 Cham, Switzerland

For Catherine and Ariadne

Acknowledgements

This book was written during the COVID pandemic of 2020. Indeed, the bulk of it was written during a national lockdown in Belgium. As I write this preface (September 2021), my hope is that we will soon see an end to all lockdowns and our previous lives can return to some sort of normality.

This book could not have been written without the love and support of my wife Catherine and our daughter Ariadne. They have both been my shining lights in this otherwise dark time. I owe each of them more than I can ever say. Thanks also to my mum and dad, Glenda and Victor, and my two brothers, Christopher and Alexander. Family and friends are everything. But this is especially true when one is living amidst a global pandemic.

Many thanks to those who gave their time and effort to review earlier drafts of this book. First, thanks to Phil Hutchinson who provided commentary on an earlier draft of this book. Second, thanks to Ludger van Dijk, who helped clarify my thinking on some of the topics discussed in this book. Third, thanks to Erik Myin, who read and commented on some parts of an earlier draft of this book. Fourth, thanks to the series editor at Palgrave who approved the manuscript for this book and to the other editorial staff who helped in the preparation of the manuscript. In particular, thanks to Brendan George who was my main point of contact at Palgrave and who kindly agreed to extend the deadline for submission of the manuscript. Fifth, thanks to two anonymous reviewers who reviewed the proposal for this book. Finally, thanks to my colleagues (both past and present) at the Centre for Philosophical Psychology at the University of Antwerp, Belgium, whose conversations have helped spur on my thinking over the years.

Some of the material that appears in this book is based on published papers. Thanks to the editors, presses and copyright holders for permission to reproduce passages from the following pieces:

Extended Mind. Appears in: V. P. Glaveanu (ed.), *The Palgrave Encyclopedia of the Possible.*

Extended Mind, Extended Conscious Mind, Enactivism. Doctoral thesis for the University of Antwerp, Belgium.

Deep Impact: Wittgenstein's enduring enactivist legacy. To appear in C. Sandis & D. Moyal-Sharrock (eds.), *Extending Hinge Epistemology*, Anthem Press, 2022.

Why enactivists should care about Wittgenstein. *Philosophia*, https://doi.org/10.1007/s11406-020-00286-3.

Why Wittgenstein was not a therapist. To be published by *Aufklärung*.

Thanks also to the Research Foundation Flanders (FWO): Project Removing the Mind from the Head: A Wittgensteinian perspective, 1209616N, which provided the financial support that made this book a reality.

All the passages by Wittgenstein reproduced in this book were done so with the kind permission of the Master and Fellows of Trinity College, Cambridge. I particularly thank Dr Bell who helped provide this permission.

Contents

1

Introduction

1.1 Thinking with and Through Wittgenstein

It has long been a commonplace, at least within philosophy departments, to claim that explaining all intelligent activity, in both humans and animals alike, required appealing to something located inside the head, where this thing is understood to be an internal representation. Sterelny summarizes this claim as follows:

> There can be no informational sensitivity without representation. There can be no flexible and adaptive response to the world without representation. To learn about the world, and to use what we learn to act in new ways, we must be able to represent the world, our goals and options. Furthermore we must make appropriate inferences from those representations. (Sterelny, 1990, p. 21)

Ramsey (2007) has pointed out that the concept of representation "is the most important explanatory posit today in cognitive research" (ibid., p. xi), and while, continues Ramsey, there are many different types of representational theories, such theories "all share the core assumption that mental processes involve content-bearing internal states and that a

V. Loughlin, *4E Cognitive Science and Wittgenstein*, New Directions in Philosophy and Cognitive Science, https://doi.org/10.1007/978-3-030-89463-4_1

correct accounting of those processes must invoke structures that serve to stand for something else" (ibid.). Following Ramsey then, I will define this core assumption as follows:

> **The Representational Orthodoxy:** A bodily internal process (in the brain, for example) represents some feature of the environment when that process stands in some relationship to that environmental feature, such that this process specifies, describes, characterizes or portrays that environmental feature. This process can be understood to be the vehicle and the way that this process specifies etc., the environmental feature can be understood to be the content. This content is truth evaluable, which is to say that the specification, description, characterization, portrayal can be evaluated as being true or false.

I have called this assumption the Representational Orthodoxy because, according to Wheeler, "the principle of representationalism is [...] one of the foundational assumptions of orthodox cognitive science" (Wheeler, 2005, p. 57). In other words, it is simply taken as given that explaining the human mind always requires invoking internal representations.

There have of course been challengers to this Orthodoxy.[1] To pick two recent examples within cognitive science, behaviour-based robotics, as pioneered by Rodney Brooks (1991), demonstrated that it was possible to build robots that were capable of performing simple tasks despite those robots having no detailed, internal knowledge of the environments in which they were operating. On the other hand, work by Ballard et al. (1997) demonstrated that subjects could exploit features of their bodies and features of their environment in order to minimize reliance on internal processing. Both strands of research have had and continue to have a significant impact on current thinking about the mind (e.g., see Hutto & Myin, 2013, p. 41).

Today such challengers to the Representational Orthodoxy have come to be grouped together under the banner of **4E** where the **E** stands for: Embodied Mind, Embedded Mind, Enacted and Extended Mind.[2] I will define the various Es within 4E as follows:

Embodied mind: cognitive processes can, on occasion, be partially constituted by aspects of our bodies.
Embedded mind: cognitive processes can, on occasion, causally depend, in some essential fashion, on features in the environment.
Enacted mind: mind and experience can be constituted by the interaction between an individual and their environment.
Extended mind: cognitive processes can, on occasion, extend to include environmental objects or processes.

There is a noticeable lack of consensus among the approaches within 4E.[3] For whereas an approach like Extended Mind can be read as compatible with the Representational Orthodoxy, Enacted Mind can be read as challenging that Orthodoxy. This lack of consensus then generates the following question, namely to what extent do 4E approaches actually challenge the Representational Orthodoxy? Answering this question matters. Consider that if 4E is to mark a genuine break from the Orthodoxy and thus as heralding something new in our thinking about the mind, then the anti-representationalist credentials of 4E need to be placed front and centre. But if so, then resolving the lack of consensus within 4E in favour of anti-representationalism turns out to be crucial. This then is the goal of this book, namely to present an anti-representationalist version of 4E.

The philosopher I will use in order to help me achieve this goal is the Austrian-British philosopher Ludwig Wittgenstein. Wittgenstein is needed, so I claim, if 4E is to become truly anti-representationalist.

For example, within Enacted Mind, two leading challengers to the Representational Orthodoxy are Radical Enactive or Embodied Cognition (REC) and Sensorimotor Enactivism. Both approaches encounter a Representationalist Objection, which they can avoid by embracing diachronic constitution. However, this commitment to diachronic constitution also involves a commitment to a Process Model, that is, a Model that seeks to picture mental phenomena in terms of processes. Wittgensteinian considerations clarify why we should reject that Model. Yet once this Model is rejected, then Enacted Mind is again faced with the Representationalist Objection. In which case, if Enacted Mind is thus to challenge the Orthodoxy, then it needs to avoid this Objection without endorsing diachronic constitution. I show how this could be done.

I present a new framework called Structural Enacted or Extended Mind (STEEM). Like other Enacted Mind approaches, STEEM identifies experiences and other mental episodes with actions. However, unlike other Enacted Mind approaches, STEEM clarifies where such identities come from, namely they come from the fulfilment of behavioural criteria. Moreover, also unlike other Enacted Mind approaches, STEEM embraces a version of Extended Mind, according to which the subpersonal processes that underpin experiences and other mental episodes can on occasion extend to include objects in the environment.[4]

STEEM thus provides an answer to our previous question. For if STEEM is accepted, then both Enacted Mind and Extended Mind challenge the Representational Orthodoxy. In which case, STEEM clarifies the anti-representationalist credentials of these 4E approaches. 4E now becomes 2E, of course. Yet this is a 2E that arguably fulfils the promise of 4E, namely to deliver a new beginning when it comes to thinking about the mind.[5]

Some may baulk at my use of Wittgenstein here. After all, Wittgenstein was well known for being opposed to developing new theories or explanations. How then can I use some of his remarks in order to develop a new framework for thinking about the mind? Such use may strike some aficionados of Wittgenstein as anathema.

However, the question (to what extent do 4E approaches actually challenge the Representational Orthodoxy?) was not one that Wittgenstein ever considered. I take this fact as licensing me to both think with and think through Wittgenstein. “Think with” in the sense that I aim to do justice to some of the themes Wittgenstein explored in his work. For example, how we need to look at behavioural criteria if we are to understand the intimate link between language and action. “Think through” in the sense that I then attempt to apply those same themes to a question that Wittgenstein did not (and might not even have wanted to) consider. The payoff is that we can then demonstrate how Wittgenstein can help shine new light on a cutting edge philosophical issue. This, so I claim, can and should be regarded as a substantial gain.

Before continuing, a brief guide for readers. This book is aimed at two distinct audiences: those with a background in 4E Cognitive Science and those with an interest in Wittgenstein. Given this diverse readership, the following instruction might help. For those with a background in 4E Cognitive Science, Chaps. 2 and 3 tread some familiar ground. As such, they can probably skip those chapters and focus instead on Chaps. 4 and 5, where I develop the Wittgensteinian considerations at the heart of this book. Conversely, for those with an interest in Wittgenstein, Chaps. 4 and 5 tread some familiar ground. As such, they can probably skip Chaps. 4 and 5 and focus instead on Chaps. 2 and 3, where I go into detail about Extended Mind and Enacted Mind respectively. Chapter 6, on the other hand, should be of interest to both groups of readers. In this final chapter, I show how Wittgensteinian considerations support a new theoretical framework (STEEM), which in turn highlights the anti-representationalist credentials of both Enacted Mind and Extended Mind. Of course, for those with no such familiarity with either 4E or Wittgenstein or for those just interested in how the whole thing hangs together, I encourage them to read all of the above chapters.

In the rest of this introduction then, I provide a detailed summary of the main topics discussed in each chapter.

1.2 Summaries of the Chapters

In Chap. 2, I consider a leading 4E approach that can be read as compatible with the Representational Orthodoxy. This 4E approach is Extended Mind. I begin by examining what can be regarded as what remains the canonical formulation of Extended Mind, namely that provided by Clark and Chalmers (1998). I focus on Clark and Chalmers' Parity Principle. This Principle states that objects can be part of the mind when those objects fulfil roles that could have been fulfilled by processes inside the head. The core idea is that when there is parity between the roles played by these objects and inner processes, then an individual's cognitive processes can be viewed as extending to include the use of that object. I claim that this principle means that Extended Mind is compatible with the Representational Orthodoxy. Extended Mind can then be read as the claim that the boundaries of internal representations, under conditions of parity, can extend to include the use of bodily external objects. Following Sutton (2010), I refer to this Parity-driven reading of Extended Mind as First Wave Extended Mind.

However, this commitment to a Parity Principle raises a question for First Wavers, namely how do they specify the roles of the environmental outer and the biological inner? Consider that if this specification is too narrow, then physical differences between environmental objects and processes inside the head will exclude the application of the Parity Principle. On the other hand, if this specification is too liberal, then potentially any object could now have parity with what is inside the head. Given the centrality of the Parity Principle to First Wave, it seems that the second option is the only one available to First Wavers.

These worries about parity then motivate looking for an alternative way to read Extended Mind. Sutton (2010) provides such an alternative. He outlines a reading of Extended Mind based on a Complementarity Principle. On this approach, objects in the environment don't need to fulfil equivalent roles to processes in the head. Rather objects in the environment can complement processes in the head within some temporary or enduring coupled individual-object system. It is thus the distinctive coupling that can occur between the individual and an object that supports talk of cognitive extension. I illustrate this by borrowing Sutton's example of medieval actors working in Shakespeare's Globe theatre. This example highlights why objects need not have parity with processes inside the head. For when these actors were coupled with the props in the Globe theatre, such coupling had a transformative effect on the actor's cognitive processes. I refer to this Complementarity reading of Extended Mind as Second Wave Extended Mind.

Yet Second Wave, just like First Wave, encounters a problem of its own. For why should the coupling between an individual and an object be sufficient to make that object part of an individual's cognitive processes? I argue that Second Wave (at least as formulated by Sutton) has no clear answer to this question. One option then is to downgrade Second Wave from an ontological claim about the boundaries of the mind to a methodological claim about how we should investigate the mind. This would avoid the previous problem. However, it would entail that First

Wave now has precedence over Second Wave, since First Wave is an ontological claim. If we thus adopt First Wave as our standard reading of Extended Mind, then Extended Mind is compatible with the Representational Orthodoxy. (Note that in Chap. 6, I consider and reject a Third Wave reading of Extended Mind.)

In Chap. 3, I consider another leading 4E approach. This approach is Enacted Mind. Enacted Mind, in contrast to Extended Mind, can be read as challenging the Representational Orthodoxy. I focus here on two leading enactivist views. These are Radical Enactive or Embodied Cognition (REC) and Sensorimotor Enactivism. I begin by considering REC.

One of the selling points of the Representational Orthodoxy is that it is alleged to offer a scientifically tractable way to investigate the mind. Recall the distinction between vehicles and contents. If we want to know what happens inside someone's head whenever they think, and we subscribe to the Orthodoxy, then we need to find the vehicle for a person's thought and this vehicle can be a brain process or processes. And if vehicles are scientifically tractable, then the same point must apply to contents. That is, it must be possible to give a credible account of how some brain process or processes can specify, describe, characterize or portray some environmental feature.

However, proponents of REC—RECers—deny that it is possible to give such an account. They label this the Hard Problem of Content (Hutto & Myin, 2013). Moreover, they reject the idea that there are any internal mental contents. They then offer their own position, according to which there are no contents or vehicles outside of shared social practices. RECers thus challenge the Orthodox assumption that explaining the human mind always requires appealing to internal representations.

Sensorimotor Enactivists (O'Regan & Noë, 2001) also challenge the Representational Orthodoxy. However, they focus on visual perception and argue that perceivers don't need to represent objects in order to perceive them. Rather perceivers need to gain embodied know-how of the sensorimotor dependencies enacted as and when perceivers have engaged with or are engaging with an object or objects. However, as I point out, it is not clear what is the role of such know-how within perception. Must a perceiver interact with the environment now in order to perceive? Or is it instead the perceiver's (perhaps long acquired) skills and abilities to move and interact with an environment that realizes their perception? In Chap. 6, I offer a possible solution to this problem.

I then consider how those keen on the Representational Orthodoxy could respond to these various enactivist challenges. For example, defenders of the Orthodoxy could invoke what I call a Representationalist Objection (RO). According to RO, we can distinguish between causal relations and constitution and insist that causal relations are atemporal whereas constitution is synchronic. This distinction, if accepted, creates a problem for enactivists. For both REC and Sensorimotor Enactivism motivate their respective claims about the mind by appealing to, among other things, temporally extended interactions between individuals and their environments. Yet if RO is correct, then while these enactivist accounts may reveal the causes of the mind, they nonetheless fail to reveal what constitutes the mind. Moreover, the only items that could then constitute the mind, according to RO, seem to be internal representations. In which case, neither REC nor Sensorimotor Enactivism challenges the Orthodoxy.

However, I argue that both RECers and Sensorimotor Enactivists can avoid RO. This requires endorsing a diachronic (as opposed to synchronic) view of constitution. According to this diachronic view, mental processes are constituted by densely coupled reciprocal causal relations. Causal relations consequently can substantiate constitutive claims about the processes that compose the mind. I then show how both REC and Sensorimotor Enactivism could be read as endorsing diachronic constitution. If so, then both versions of Enacted Mind can be read as avoiding RO. If they can avoid RO, then both REC and Sensorimotor Enactivism can maintain their respective challenges to the Representational Orthodoxy.

I then move from discussion of 4E to discussion about Wittgenstein. I begin Chap. 4 by considering a position within the vast secondary literature on Wittgenstein's remarks on rule following. This position is that of Baker and Hacker (2009). I also consider Macha's (2015) support and development of Baker and Hacker's position. For my purposes, I assume that these authors are correct in their assessment of what Wittgenstein says about following rules.

According to Baker and Hacker, there is an internal relation between a rule and what accords with that rule. Using Macha's terminology, I understand this to mean that there is an internal relation between the concept of a rule and the concept of what accords with it. Moreover, we can demarcate the behavioural criteria needed to follow a rule from how one learns a rule. These ideas are related. For if what makes a rule a rule is the internal relation between the concept of the rule and the concept of what accords with it, then the behavioural criteria needed to follow a rule will involve the enactment of this relation and this enactment will be distinct from how one learned the rule. As I clarify in Chap. 4 and subsequent chapters, I use the term "enacting an internal relation" as a philosophical term of art, that is, as a way of focusing on the behavioural criteria that must be fulfilled in order to be a rule follower.

I then use this analysis to show why we should reject what I call a Process Model. According to the Process Model, what is true about teaching and learning a rule is also true about the behavioural criteria for following a rule. That is, just as teaching and learning a rule is a sequence of actions spread out in time and so is a process, then so too is following the rule. However, once we accept that the behavioural criteria for following a rule involve enacting an internal relation between the concept of a rule and the concept of what accords with it, which ensures that the behavioural criteria needed to follow a rule are distinct from how one learns a rule, then even if teaching and learning a rule is indeed a process, the same is not true of following a rule. Given that internal relations are atemporal and processes are atemporal, then internal relations cannot be understood as processes. If so, then rule following cannot be modelled in terms of processes. We should thus reject the Process Model. This rejection of the Process Model has significant implications for 4E Cognitive Science (see Chap. 6). I will thus spend this chapter and the next developing these implications in some detail.

In Chap. 5, I show that the analysis developed in Chap. 4 is not restricted to following rules but can also be applied to our talk about the mind. In particular, I show that rejecting this Model allows us to dissolve a problem that our talk of the mind may seem to pose.

Consider the many statements we make about our minds. These statements can seem to raise a problem. For when we talk about our minds, it can seem natural to suppose that something inside us (usually taken to be inside our heads) must have prompted us to make those statements. We can then view such internal promptings as being processes occurring inside our brains. However, when we examine those statements, we usually find that they don't in fact reference anything inside our heads. Moreover, we rarely simply talk about our minds. More often than not, we instead act out such talk via what we say and do. But if so, then talk about our minds seems to both direct us inwards towards the brain and outwards towards verbal and non-verbal bodily behaviours. How then are to make sense of our talk about the mind?

In order to tackle this problem, I first discuss pain experience. For example, what happens when a child falls over and turns to their parents for comfort? Does the parent's comforting make the child's pain dissipate or go away? Using Wittgensteinian considerations, I distinguish between natural pain behaviours and new pain behaviours. New pain behaviours, for example, saying "foot sore" while pointing at one's foot, require mastery of a concept, since they involve the enactment of an internal relation. If so, then new pain behaviours require that we reject the Process Model, since internal relations are atemporal whereas processes are atemporal. Once we reject this Model, then we can answer the previous question. The parent's comforting of the child neither dissipates the child's pain nor makes the child's pain go away. Rather the child seeks comfort when they fall over and parents provide that comfort because these behaviours are part of a child and parents' natural pain behaviours. These behaviours are meaningful. Nonetheless, not all pain behaviours are the same. Some are bodily and have no criteria. Others are expressed through language and so do have criteria.

Go back then to our original problem. What gives sense to our statements about our minds are actions that involve internal relations, for example, internal relations between statements or internal relations between statements and bodily behaviours. That is, it is the fulfilment of these behavioural criteria that gives these statements and their accompanying behaviours their meaning. This then allows us to dissolve our initial problem. Talk about our minds only seems to pose a problem because we assume that something must happen when we make our various statements about our minds. Moreover, what happens then must be a set of processes occurring inside the brain. This assumption is an example of the Process Model. Yet if what gives sense to our statements about our minds are actions that involve internal relations and internal relations are atemporal, then these statements cannot be understood in terms of processes. We should thus reject the Process Model. Once we reject this Model, then there is no problem when it comes to our talk about our minds. Our statements about our minds have meaning because of the contexts within which we make those statements.

I now move the discussion from Wittgenstein back to 4E. Recall that in Chap. 2, I showed that First Wave Extended Mind was compatible with the Representational Orthodoxy. In Chap. 3, I showed that Enacted Mind—both Radical Enactive or Embodied Mind (REC) and Sensorimotor Enactivism—challenged the Representational Orthodoxy. Given that both Extended Mind and Enacted Mind are part of the wider 4E family, then it follows that 4E offers a conflicting response to that Orthodoxy. The aim of Chap. 6 then is to show how 4Eers can resolve this conflict and so build a consensus against that Orthodoxy.

I first identify the two conditions both Enacted Mind and Extended Mind must fulfil if they are to challenge the Representational Orthodoxy. First, they must avoid the Representationalist Objection (RO). Second, they must avoid RO without endorsing diachronic constitution. This is because diachronic constitution is linked to the Process Model. The analyses given in Chaps. 4 and 5 showed why we should reject that Model. But if so, then we should also reject diachronic constitution.

I then outline a theoretical framework that, so I claim, can fulfil these two conditions. I call this framework Structural Enacted or Embodied Mind or STEEM. Given that STEEM can fulfil both conditions, then STEEM challenges the Representational Orthodoxy and so demonstrates the anti-representationalist credentials of both Enacted Mind and Extended Mind.

For example, both Sensorimotor Enactivists and RECers aim to dissolve the alleged explanatory gap between experience and action by identifying experiences with actions. Yet while this identification dissolves any such gap, it raises the following question, namely where do these identities come from? I answer this question by appealing to Structural Enacted Mind. According to Structural Enacted Mind, identities between action and phenomenal experience come from the fulfilment of behavioural criteria. This is because the behavioural criteria needed for mastery of phenomenal concepts involves enacting internal relations. These relations are definitional of phenomenal experiences. We can thus identify the enactment of an internal relation with a phenomenal experience. If so, then experiences are structural. For example, the feeling of softness is what it is due to the fulfilment of the criteria needed for mastery of the concept of softness, which involves enacting an internal relation.

If Structural Enacted Mind is correct, then we should reject the Process Model. There are no processes that occur when someone squeezes, say, a sponge, which ensure that the individual experiences the feeling of softness. Rather the individual experiences the feeling of softness due to the internal relations involved. As I show, Structural Enacted Mind avoids RO without endorsing diachronic constitution.

I then consider Structural Extended Mind. As Chap. 2 illustrated, we can distinguish between a First Wave and Second Wave Extended Mind. Chapter 2 also illustrated that while both readings face problems, priority could be given to First Wave. However, there is also a possible Third Wave Extended Mind. Third Wave, unlike First Wave, rejects the Parity Principle and so is not compatible with the Representational Orthodoxy. However, Third Wave, at least on one reading, appeals to diachronic constitution (e.g., Kirchhoff & Kiverstein, 2018). Given the need to reject diachronic constitution due to its link with the Process Model, then this gives us grounds to reject Third Wave. But if so, then it follows that both First, Second and Third Wave should be rejected.

I propose that we instead appeal to the distinction between the Hypothesis of Extended Mind (HEM) and the Hypothesis of Extended Cognition (HEC). HEM is a version of First Wave and so should be rejected. This leaves HEC. I show that HEC is compatible with Structural Extended Mind. Accordingly, experiences (and other mental episodes) can be identified with actions. Nonetheless, the sub-personal processes that underpin those actions can be understood as sometimes extending to include our use of objects. Focusing on the example of remembering, I illustrate how this would work. I show that remembering a text can be understood to be an action that involves internal relations. Yet the processes that underpin that action can, under certain circumstances, extend to include lines or marks made on a page. This example is thus an instance of Structural Extended Mind. Crucially, this example avoids RO without endorsing diachronic constitution.

A STEEM framework consequently ensures that both Enacted Mind and Extended Mind can unify their response to the Representational Orthodoxy. 4E now becomes 2E. STEEM thus demonstrates how a 2E framework can offer a new approach to thinking about the mind.

Notes

1. These challengers include pragmatists, phenomenologists, ecological psychologists and ethnomethodologists.
2. According to Mark Rowlands, it was Shaun Gallagher who invented the term 4E. See Rowlands (2010, p. 219, footnote 4).
3. The idea that there is a lack of consensus within 4E is well understood. For example, Thompson (2018) writes: "Some [4E] approaches try to revise traditional models of computation, mental representation, and agency by grounding them on functional notions of the physical body; other approaches depart from the computational tradition by rejecting its core assumptions about the mind".
4. Hutto and Myin (2013, p. 7) reject Extended Mind on the grounds that it fails to rule out the idea that the mind is, by default, brain-bound and so only occasionally extends. STEEM embraces a version of Extended Mind (see Chap. 6, Sect. 6.4). This distinguishes STEEM from Hutto and Myins' own enactivist proposal.
5. I won't have much more to say about either Embodied or Embedded mind in this book. There are two reasons for this. First, I take Embodied Mind to be assumed by both Enacted and Extended Mind. For example, even representationalist friendly versions of Extended Mind, like First Wave, recognize the importance of the body (see Chap. 2, Sect. 2.3). Second, it is only Embedded Mind that conceivably challenges Extended Mind. Nonetheless, I think Extended Mind can withstand this challenge (see Chap. 2, Sect. 2.2 where I defend this point).

References

Baker, G., & Hacker, P. (2009). *Wittgenstein: Rules, Grammar and Necessity: Volume 2 of an Analytical Commentary on the Philosophical Investigations, Essays and Exegesis 185–242* (2nd ed.). Wiley Blackwell.

Ballard, D., Haynoe, M., Pook, P., & Rao, R. (1997). Deitic Codes for the Embodiment of Cognition. *Behavioural and Brain Sciences, 20*, 723–767.

Brooks, R. (1991). Intelligence Without Representation. *Artificial Intelligence, 47*, 139–159.

Clark, A., & Chalmers, D. (1998). The Extended Mind. *Analysis, 58*(1), 7–19.

Hutto, D., & Myin, E. (2013). *Radicalising Enactivism: Basic Minds Without Content*. MIT Press.

Kirchhoff, M., & Kiverstein, J. (2018). *Extended Consciousness and Predictive Processing: A Third Wave View*. Routledge.

Macha, J. (2015). *Wittgenstein on Internal and External Relations: Tracing All the Connections*. Bloomsbury Academic.

O'Regan, K., & Noë, A. (2001). A Sensorimotor Account of Vision and Visual Consciousness. *Behavioural and Brain Sciences, 24*(5), 939–1031.

Ramsey, W. (2007). *Representation Reconsidered*. Cambridge University Press.

Rowlands, M. (2010). *The New Science of the Mind: From Extended Mind to Embodied Phenomenology*. MIT Press.

Sterelny, K. (1990). *The Representational Theory of Mind: An Introduction*. Blackwell.

Sutton, J. (2010). Exograms and Interdisciplinarity: History, the Extended Mind, and the Civilising Process. In R. Menary (Ed.), *The Extended Mind* (pp. 189–226). The MIT Press.

Thompson, E. (2018). Review of: Evolving Enactivism: Basic Minds Meets Content. *Notre Dame Philosophical Reviews*.

Wheeler, M. (2005). *Reconstructing the Cognitive World: The Next Step*. MIT Press.

2

First and Second Wave Extended Mind

2.1 Introduction

In the Introduction, I outlined what I called the Representational Orthodoxy, which is the assumption that explaining the human mind always requires invoking internal representations. I then pointed out that there is lack of consensus within 4E as to whether or not 4E challenges the Representational Orthodoxy. In this chapter (and the next), I will offer evidence to support this claim.

I start then with Extended Mind. In particular, I will focus on a reading of Extended Mind that can be regarded as compatible with the Representational Orthodoxy.

2.2 First Wave Extended Mind

Most of us have at some point played (or at least have heard about) the computer game Tetris. For those that haven't, the goal of this game is to alter the shapes of falling objects so that these objects can form a solid line at the bottom of a screen. Players can do this in two ways. They can either

V. Loughlin, *4E Cognitive Science and Wittgenstein*, New Directions in Philosophy and Cognitive Science, https://doi.org/10.1007/978-3-030-89463-4_2

quickly calculate in their heads how these objects should be altered before then making the necessary adjustments on the screen. Or they can instead use a joystick attached to the computer to make those same adjustments.

The philosophers Andy Clark and David Chalmers (1998) have argued that even though these two methods for manipulating the falling objects are distinct, they can nonetheless be understood as equivalent. This is because objects can be part of the mind when those objects fulfil roles that could have been fulfilled by processes inside the head. In subsequent debates over what has been called Extended Mind, this move by Clark and Chalmers has been referred to as the Parity Principle (PP) (Menary, 2010a). The basic idea behind PP then is that, under certain circumstances, the physical location of an object need not exclude it from being part of a cognitive process. For what matters is not the physical location of an object but rather how an individual's use of that object contributes to their performance of a given task. PP is thus, in Clark's (2008a) words, a "veil of metabolic ignorance" or an insistence on "equality of opportunity [or contribution]: avoiding a rush to judgment based on spatial location alone" (ibid., p. 114). And where there is parity of contribution between use of an object and in-the-head cognitive processing, there can also be parity of cognitive status (Wheeler, 2010, p. 248).

However, this parity of cognitive status requires that use of an object and in-the-head cognitive processing be understood in terms what is called their functional role. Functionalism, broadly understood, separates functional role from physical materiality in order to argue that what makes a process cognitive is not whatever physical substance composes that process but rather the role that process plays within a larger causal network of inputs and outputs (Levin, 2008). Once functionalism is accepted, then the physical details of an object and how it is used only matter to the extent that these details help the object fulfil the requisite functional role.

Go back then to the Tetris example. If the above is accepted, then how the player grasps the joystick, positions their body and/or moves when using the joystick is only important to the extent that these actions fulfil a functional role. Once the player's use of the joystick is framed in these functional terms, then there can be parity or equivalence between, on the one hand, the player's use of a joystick, and on the other, the player's manipulation of the shapes of the falling objects in their head. This is because the use of the joystick can be understood as part of a process that were that process to occur inside the player's head we would regard it as a cognitive process. Given this parity or equivalence, then the player's use of the joystick can be understood as extending the cognitive processing of the player.

However, perhaps the most famous example of Extended Mind comes not from consideration of the computer game Tetris, but rather from the following thought experiment outlined by Clark and Chalmers, namely the thought experiment of Otto and Inga.

Otto and Inga are two residents of New York. Otto is in the early stages of Alzheimer's disease and so has begun to have real difficulty in remembering things. He tries to overcome this difficulty by always carrying with him a notebook. Whenever Otto hears something that he thinks might be important and so worth remembering, he quickly writes it down in his notebook.[1] However, Otto's friend Inga has no such difficulties with her memory. Her ability to remember remains fully intact.

Consider then the following scenario. Inga hears about an exhibition at the Museum of Modern Art (MoMA), which she really wants to see. She thinks for a moment and recalls that MoMA is on 53rd Street. Later that same day she then visits the museum. Otto also hears about the same exhibition as Inga and, like Inga, he decides he really wants to see it. However, rather than trying to recall the location of the museum, Otto simply opens up his notebook and finds where he has written down the location. Later that same day Otto then visits the museum.

According to Clark and Chalmers, both Inga and Otto have an unconscious belief about the location of the museum. The only difference is that whereas Inga's unconscious belief is located inside her head, Otto's unconscious belief is partially located outside his head in his notebook. As we have seen, PP means that the physical location of an object is not functionally relevant. What is functionally relevant is instead the role the object plays in an individual's performance of a task. This applies to Otto. For Otto does with his notebook what Inga does with her biological memory. But if so, then Otto's use of his notebook and Inga's use of her memory can be understood as being on a par, since the notebook is part of a process that were that process to occur inside the head, as it does in the case of Inga, then we would regard it as a cognitive process. In which case, Otto's unconscious belief about the location of the museum can be understood as extending to include his use of the notebook. Otto is thus an example of Extended Mind.[2]

Clark and Chalmers regard Otto's unconscious belief as a mental state. Yet there are grounds to think that the Otto-Inga thought experiment should instead be framed in terms of processes rather than states. For both Inga and Otto engage in a process in order to remember the location of the museum: Inga thinks for a moment whereas Otto accesses his notebook. Once both Inga and Otto are understood as engaged in a process, then it is apparent how PP can be applied and thus apparent how Otto can be an example of Extended Mind.[3] We could deny this and instead claim that it is the interaction between Otto and his notebook that makes him an example of Extended Mind. However, as I show later (see Sect. 2.4 of this chapter), this claim would be vulnerable to the coupling-constitution fallacy.

Either way, the Otto-Inga thought experiment has given birth to a lively debate within the literature on Extended Mind (see the introduction to Menary, 2010a, for an overview). For my purposes, it will suffice to note that the example of Otto and Inga, just like the Tetris example, demonstrates, so I would argue, the centrality of PP to making the case for Extended Mind. For in both examples, it is the alleged parity or equivalence between use of an environmental object (a joystick, a notebook) and in-the-head cognitive processing that then supports the claim that such use can then be part of an extended cognitive process (manipulating the shapes of falling objects on a screen, remembering the location of the museum). If so, then these examples illustrate that the fate of Extended Mind, at least as formulated by Clark and Chalmers, is closely tied to the fate of PP.

This conclusion could be disputed, however. It could be argued, for instance, that Clark and Chalmers identify a number of other conditions as needed for cognitive and mental extension. In the case of Otto, these conditions could include the fact that Otto must constantly use the notebook, have easy access to its contents, automatically endorse those contents and have done so consciously at some point in the past. It is thus the fulfilment of these four conditions that supports talk of Extended Mind. But if so, then PP does not have the central importance I have accorded it.

However, as Rupert (2004) points out, we could reject Extended Mind and instead embrace what Rupert calls the Hypothesis of Embedded Mind (HEMC) or more simply Embedded Mind. Embedded Mind recognizes that an individual's cognitive processes can involve causal dependencies between the individual and their use of an object or objects. These dependences would seem to fulfil the four conditions Clark and Chalmers identify. But if so, then reliance on these conditions alone only provides an argument for Embedded Mind and not also an argument for Extended Mind. This then supports my previous assessment, namely that the PP is indeed needed if we are to talk about Extended Mind (at least as Extended Mind is understood by Clark and Chalmers).

Nevertheless, others keen on Extended Mind have still sought to distance themselves from PP. This had led to what have been called Second Wave and Third Wave versions of Extended Mind (Sutton, 2010). Given these alternative readings, I will thus refer to the Clark and Chalmers' formulation of Extended Mind as a First Wave reading of Extended Mind or more simply First Wave Extended Mind.

Recall then the Representational Orthodoxy, that is, the assumption that explaining the human mind always requires invoking internal representations. Is PP compatible with this assumption? I will argue that it is.

PP only works if we agree to assume that in-the-head processing sets the standard for what counts as cognitive or mental. For only if this assumption is correct can we then posit parity or equivalence between what goes on inside the head and the use of bodily external objects (joysticks, notebooks) and thereby claim that these usages are part of an extended cognitive or mental process. However, as Loughlin and Zahidi (2017) have argued, Clark and Chalmers give no indication as to why what happens inside the head should be regarded as cognitive or mental. Di Paolo (2009) offers a similar conclusion. He points out that while First Wave challenges some intuitions about the boundaries of the mind, it leaves other intuitions entirely intact, such as the intuition that processes inside the head are cognitive or mental simply because they occur inside the head. If these assessments of First Wave are correct, then PP looks entirely unsupported. Given the centrality of PP to First Wave, this is a problem.

One way then for Clark and Chalmers to resolve this problem, so I claim, is to invoke internal representations. That is, bodily internal processes (in the head, say) can be regarded as cognitive or mental because these processes are representational, that is, these processes act as the vehicles for semantic content. This would have the advantage of demonstrating why what is inside the head sets the standard for what counts as cognitive or mental and so give substance to PP. The cost however is that on this reading of PP, First Wave Extended Mind becomes the claim that the boundaries of internal representations can, under conditions of parity, extend to include the use of objects. In short, the cost is that First Wave should be understood as compatible with the Representational Orthodoxy. Yet given the need to support PP, I think this is a cost that First Wavers would and should be prepared to bear.

2.3 Snakes and Adders

In his book *Supersizing the Mind*, Clark (2008a, p. 69) discusses the example of Ada. Ada is an accountant and good at her job. Part of what allows Ada to excel at her job, argues Clark, is that she doesn't try to commit large numbers to memory whenever she has to work through some difficult accounting problem. Instead, Ada simply deploys what are called minimal memory strategies, that is, she uses pen-and-paper (and perhaps other devices) in order to minimize the demands she needs to make on her own biological memory. So, just as you or I may leave a note on our refrigerator door in order to remind ourselves to buy milk, Ada always makes notes about the numbers she is scanning and then moves backwards and forwards between her quickly scribbled notes on a "just-in-time, need-to-know basis, briefly shunting specific items of information into and out of short-term bio-memory" (ibid.). By such means, Ada thereby simplifies a complex problem and so can solve it in a ready manner. For all Ada now needs to do, according to Clark, is to follow the "trails through the scribbled numbers, relying on self-created external traces every time an intermediate result is obtained" (ibid.).

Think back then to the Parity Principle (PP). PP seems applicable to Ada. For Ada's interactions with her pen-and-paper can be part of a process that were that process to occur inside her head we would regard that process as cognitive. If so, then Ada's use of the pen-and-paper can be understood as extending her in-the-head cognitive processing. Ada is thus an example of First Wave Extended Mind.

Note that the Ada example, just like the Tetris example and Inga-Otto example discussed previously, takes as given functionalism, that is, it takes as given the idea that cognitive processes are to be understood solely in terms of their functional roles. Clark agrees: "it is the roles played by various elements, and not the specific ways those elements are realized, that do the explanatory work" (ibid., p. 14). As discussed earlier, this means that the materiality of, for example, joysticks, notebooks or pen-and-paper only matters to the extent that these devices or objects help the individual perform some task. This commitment to functional role leads Clark to invoke what he calls platform neutrality (ibid., p. 201). A process is platform independent, insists Clark, if the functional profile for that process, that is, the causal network of inputs and outputs that realize that process, can be implemented on more than one platform. In which case, the process can be said to be platform independent (ibid., p. 198).

Clark illustrates this idea of platform independence using his science fiction example of Adder, a snake-like creature that is placed on top of a touch screen (ibid., p. 203). By wriggling its body in a particular way, Adder can get symbolic tokens to appear on this screen, which Adder then processes using a form of Braille. Clark claims that Adder could "carry out the same complex accounting as the standard, pen-and-paper accountant Ada […] [since] [e]ach [Ada and Adder] implements the same extended computational process. They even, we may suppose, divide the biological and nonbiological contributions in the same way, making use of external storage and notations at exactly the same points in their distributed problem-solving routines" (ibid.).

If Adder can instantiate the same distributed combination of memory, actions, storage and perceptual access as Ada, then the process implemented by both Ada and Adder can be understood to be platform independent. For whereas Ada implements that process via her use of bodily internal and bodily external resources, Adder implements that same process via bodily external resources alone.

However, Clark seeks to reconcile this platform independence with recognition of the, in his words, "profound contributions that embodiment and environmental embedding make to the solution of [problems]" (ibid., p. 202). That is, Clark wishes to recognize that the physical resources used by Ada and/or Adder are crucial to their individual performances of the accounting task. Consider how Ada continuously scans back and forth between her scribbled notes and the columns of numbers in order to solve the accounting problem. Or consider how Adder must wriggle its entire body on the touch screen in order for symbolic tokens to appear, which it then reads using a form of Braille.

Yet, as Clark acknowledges (ibid., p. 204), there is an obvious tension here. For if bodily interactions play such a significant role in both examples, then the process involved is, contrary to what was claimed above, bound to a particular platform and so not platform independent. Clark aims to resolve this tension, however, by insisting that the body should only be viewed as "the gateway to intelligent offloading", or "a bridging instrument enabling repeated emergence of new kinds of distributed information processing organization" (ibid., p. 207). In other words, what matters is how Ada and Adders' bodily actions help them store, transmit and transfer information. In which case, while bodies matter, they only matter to the extent that they can fulfil the required information processing roles.

This, insists Clark, is compatible with platform independence. For while Ada and Adder obviously have different bodies and use those bodies in distinct ways, both Ada's body and Adder's body can fulfil the same information processing roles in their individual performances of the task. In which case, platform independence is compatible with recognizing the crucial contributions different bodies can make to performance of a task.

However, notice that platform independence, at least as it is understood above, involves no commitment to the location of the process. For whereas this process extends in the case of Ada outside her head to include her use of pen-and-paper, there is no such extension in the case of Adder, since Adder's processing is entirely realized via it's bodily external wriggling on the touch screen. This shouldn't be a problem, since, as follows from PP, the physical location of an object is not functionally relevant. Still, the clear bodily differences between the two examples mean that if PP is to apply to Ada and Adder, then everything hangs on how we specify the functional roles of memory, actions, storage and perceptual access involved in these two examples.

One way to do this, so it seems, is to set the parameter that specifies the functional role in a liberal way. Yet this raises the problem of what is called cognitive bloat. For then all manner of what we might intuitively regard as non-cognitive systems could potentially fulfil this parameter and thus would have to be regarded as cognitive. Sprevak (2009) develops this point using what he calls the Martian Intuition. The Martian Intuition is the idea that if we can imagine an individual's use of an object as part of a process that could occur inside the head of a Martian, then that process is cognitive, since the process now has parity with the cognitive process occurring inside the head of the Martian. The problem is that this leads to cognitive bloat, since any and every process in the environment could potentially occur inside the head of a Martian (it is after all a Martian) and so could be regarded as cognitive.

We can avoid this explosion in cognitive possibilities, however, by instead restricting the aforementioned parameter using a fine-grained setting. This would have the advantage of recognizing the importance of bodily details. Yet it would do so at the cost of PP. For on this strict setting, bodily differences between Ada and Adder will then exclude regarding Adder's bodily wriggling as having parity or equivalence with Ada's use of pen-and-paper.

The larger worry here is that it is not apparent how Clark can, on the one hand, insist that there can be parity between in-the-head cognitive processing and use of objects while also insisting, on the other hand, that such parity can recognize the importance of bodily details. It seems Clark must make a choice. Either specify functional roles in a precise manner and thereby recognize the importance of bodily details; but then this is not compatible with PP. Or instead specify functional roles in a liberal manner and so preserve PP but then accept cognitive bloat.[4] Yet given the centrality of PP to First Wave (see Sect. 2.2 of this chapter), it seems the second option is the only viable one for a First Waver like Clark.

Of course, one way for proponents of Extended Mind to avoid these thorny issues is to simply drop any commitment to PP. However, if this is to work, then such proponents need to find alternative ways to motivate claims about Extended Mind. In the following section, I will consider such an alternative.

2.4 Second Wave Extended Mind

First Wave, as we have seen, is characterized by its commitment to the idea that cognitive extension involves parity or equivalence between in-the-head cognitive or mental processing and use of environmental objects. This insistence on parity or equivalence was called the Parity Principle (PP). As I pointed out in Sect. 2.2 of this chapter, other proponents of Extended Mind have challenged PP and such challenges have led to what have been called Second Wave and Third Wave readings of Extended Mind (Sutton, 2010). Here I will consider a Second Wave reading of Extended Mind or just Second Wave Extended Mind. In particular, I will focus on Sutton's Complementarity reading of Second Wave and how this reading separates itself from any commitment to PP.[5]

As discussed in Sect. 2.3 of this chapter, Clark attempts to reconcile what he calls platform independence with an acknowledgement of the importance of the role of the body. However, Sutton argues that any such commitment to platform independence must ignore what is in fact attractive about Extended Mind, which is both recognition of the different ways in which individuals can use objects to extend their cognitive processing and how some objects but not others lend themselves to such usages.

Going into more detail, Sutton reminds us how radically different individuals are when it comes to their use of objects to help them perform certain tasks (ibid., p. 199). Grasping these individual differences thus requires a focus on individual usage of objects and not on making blanket claims about human mentality, since only some individuals and not others are going to use objects in ways amenable to Extended Mind.

On the other hand, by insisting on platform independence, continues Sutton, First Wave effectively wipes clean the differences between individual objects (ibid.). After all, on First Wave reading, all that matters is that use of an object fulfils the required information processing roles (as noted in Sect. 2.3 of this chapter). Yet, insists Sutton, differences among objects may be crucial, since such differences may explain why only some objects and not others can transform an individual's cognitive processing. In which case, First Wave, according to Sutton's assessment, fails on two fronts, since it not only ignores the differences among individuals and how they use objects but also ignores differences among the objects themselves.

Sutton thus defends what can be called a Complementarity reading of Extended Mind. This Complementarity reading drops any commitment to PP, that is, it denies that objects in the environment must fulfil equivalent roles to processes in the head. Rather objects in the environment can complement processes in the head within some temporary or enduring coupled individual-object system. It is thus the distinctive coupling that can occur between the individual and an object that supports talk of cognitive extension.

Sutton illustrates his Complementarity reading by discussing Tribble's (2005) study of Shakespearean actors at the Globe theatre. As Tribble points out, such actors had to learn new lines in multiple plays in short spaces of time. Moreover, they often had to perform these tasks without access to the plays themselves. This raises an obvious question, namely how were these actors able to perform these complex tasks?

Sutton offers the following answer (Sutton, 2010, p. 203). He points out that the Globe Theatre was structured in such a way that the task of the actor was to get to a prearranged spot at a prearranged time. Once they did so, then they could quickly rehearse their lines and then perform them on stage. As such, actors weren't required to remember vast amounts of text or cues. Instead, by making use of the props in their environment, they could accomplish tasks within that environment that they could not accomplish alone or outside that environment. It was thus the complex interplay between these actors and the props available in the Globe theatre that enabled these actors to learn new lines in multiple plays in short spaces of time.

Note that, on this Complementarity reading, there is no requirement for parity or equivalence between what is inside the actor's heads and the props within the Globe theatre. Indeed, this lack of parity or equivalence turns out to be crucial. For the props in the theatre can have properties not possessed by in-the-head processing and it is this very difference between the environmental outer and the biological inner that allows these props to transform the cognitive processing of these actors. In which case, it is the coupling between actors and props that supports regarding such props as extending the cognition of these actors.

However, this reliance on coupling potentially exposes a Complementarity reading to the following problem, which Ramsey articulates as follows:

> [w]hat [second wave] authors need, but do not really provide, is an argument for treating external structures as not only important for (and integrated with) cognitive systems during various cognitive tasks…but for also treating them as actually [cognitive]. (Ramsey, 2010)

Ramsey's criticism is a version of what is often called the coupling-constitution fallacy (Adams & Aizawa, 2008, p. 68). The alleged fallacy is thinking that coupling between an individual and an object is sufficient to demonstrate that use of an object is part of those processes that compose the individual's mind. This is because no use of an object, no matter how complex or extensive, can make that object a part of an individual's mind. The problem, for Sutton at least, is that a Complementarity reading looks guilty of committing this fallacy.

Suppose, for example, that Sutton is correct to claim that the props in the Globe Theatre were such that, when coupled to the actors, those props transformed the cognitive processing of those actors. If the coupling-constitution fallacy is correct, then none of this is sufficient to demonstrate that this transformation thereby extended the cognition of these actors. For perhaps mastering the resources of this theatre simply acted as the much-needed prompt for, say, an individual actor to learn their various lines or cues. Nonetheless, the actual learning of those lines or cues still involved internal representations occurring inside the actor's head. In which case, while the Globe and its props may have scaffolded the actor's cognizing, indeed been such that these actors could perform tasks within the Globe Theatre that they could not perform alone or outside of that theatre, none of this demonstrates that such props thereby extended the cognitive processing of these actors. To think otherwise, as Sutton seemingly does, is to commit the coupling-constitution fallacy.

Of course, as noted earlier, Sutton is opposed to making blanket claims about human mentality, given his focus on differences between individuals and differences between objects. So understood, Complementarity might better be understood as a methodological claim rather than as an attempt to present some new metaphysics of the mind (Sutton, 2010, p. 191). Yet if Complementarity is restricted to a methodological claim, then First Wave and Second Wave do not have equivalent status. This is because First Wave, with its commitment to PP, is explicitly an ontological claim (see Clark, 2008b, p. 12), that is, it is a claim about how the boundaries of the mind can sometimes extend. Indeed, Second Wave would now seem to amount to nothing more than the insistence that proponents of Extended Mind do First Wave better, that is, recognize that PP will only be applicable in some situations and not others.

Moreover, it is worth noting that First Wave does not seem to display the same vulnerability to the coupling-constitution fallacy as Second Wave. Think back to Otto and his notebook. Proponents of this fallacy, such as Adams and Aizawa, might object that even if Otto relies on his notebook in order to remember the address of MoMA, such that removal of the notebook would thereby ensure that Otto could not remember the address, this only demonstrates a causal dependence between Otto and his notebook. It does not demonstrate the further claim that the processes that compose Otto's unconscious belief about the location of the museum also extend to include his use of the notebook. At best, Otto is an example of a Rupert-style Embedded Mind.

However, for First Wavers, while Otto may indeed be coupled to his notebook during those particular moments when he accesses the address of the museum, it is not coupling that extends Otto's mind. Rather it is the parity or equivalence between what Otto does with his notebook and what Inga does with her biological memory (see Roberts & Columbetti, 2015, p. 9). Of course, as noted earlier (Sect. 2.3 of this chapter), PP encounters significant problems of its own. Nonetheless, the point remains that while the coupling-constitution fallacy has clear application to Second Wave, it does not, so I would argue, have a similar clear application to First Wave.

2.5 Conclusion

In this chapter, I have shown that a First Wave reading of Extended Mind can be understood as compatible with the Representational Orthodoxy. This compatibility follows from First Wave's commitment to a Parity Principle (PP). I then considered a possible problem with this Principle, namely how to specify the functional roles of the environmental outer and biological inner. If this specification is set in a liberal manner, then this leads to cognitive bloat. Alternatively, if it is set in a precise manner, then PP no longer applies. Given the centrality of PP to First Wave, only the first option is a viable one for First Wavers.

I then moved on to discuss those proponents of Extended Mind who reject PP in favour of a Complementarity reading or Second Wave Extended Mind. Complementarity also encounters a problem. For it arguably fails to demonstrate why coupling between an individual and use of an object means that such use is thereby part of and so extends the cognitive processes of that individual. Second Wave, in other words, is vulnerable to the coupling-constitution fallacy. Second Wave can avoid this fallacy by restricting itself to a methodological claim. But then First Wave has priority over Second Wave. Moreover, First Wave does not display the same vulnerability to this fallacy as Second Wave.

In Chap. 6 (Sect. 6.4), I will discuss a Third Wave Extended Mind. For now, I claim that since First Wave has priority over Second Wave and First Wave is compatible with the Representational Orthodoxy, then Extended Mind can be read as compatible with that Orthodoxy.

Notes

1. For a real world example of Otto, consider an example discussed in Drayson and Clark (2019). They describe how a group of Alzheimer's patients living in the United States had decorated their homes with written and visual reminders, which enabled these patients to offset their memory impairments and in doing so live without constant medical care.
2. The unconscious nature of Otto's belief matters here, since, according to Clark (2009), Extended Mind does not apply to consciousness. That is,

even if objects can play an active role in the unconscious mind, the same does not hold true for the conscious mind. That is, there is no Extended Conscious Mind. In previous work, I have disputed the veracity of Clark's claim (see Loughlin, 2013, 2014). However, in Chap. 6 of this work, I will offer a reading of Extended Mind that pitches Extended Mind firmly at an unconscious or sub-personal level. If Structural Enacted or Extended Mind (STEEM) is accepted, then only the sub-personal processes that underpin minds can be regarded as extending.

3. Rowlands (2010, p. 67) seems to endorse a version of this view, which he calls process externalism.
4. Kirchhoff and Kiverstein (2018) identify an equivalent problem and note that it is a long-standing one within functionalism, for example, Block (1978). It is not clear, to me at least, how First Wavers like Clark can resolve this problem.
5. There are alternative Second Wave readings. One such example is Menary's Cognitive Integration view. For Menary (2010b, pp. 240–241), cognitive processes are hybrid processes that constitutively involve the bodily manipulation of external vehicles. However, as noted by Kirchhoff and Kiverstein (2018), Menary's Second Wave reading (in contrast to Sutton's Second Wave reading) looks closer to Enacted Mind rather than Extended Mind, given its insistence on the importance of bodily manipulation of external vehicles. Given that Menary's view can be read as a version of Enacted Mind, I have thus chosen to focus on Sutton's reading, since as I take it as a clear-cut example of Extended and not Enacted Mind.

References

Adams, F., & Aizawa, K. (2008). *The Bounds of Cognition*. Wiley-Blackwell Publishing Ltd.

Block, N. (1978). Troubles with Functionalism. *Minnesota Studies in the Philosophy of Science, 9*, 213–325.

Clark, A. (2008a). *Supersizing the Mind: Embodiment, Action and Cognitive Extension*. Oxford University Press.

Clark, A. (2008b). Pressing the Flesh: A Tension in the study of Embodied, Embedded Mind? *Philosophy and Phenomenological Research, 76*(1), 37–58.

Clark, A. (2009). Spreading the Joy: Why the Machinery of Consciousness Is Still Probably Inside Your Head. *Mind, 118*(472), 963–993.

Clark, A., & Chalmers, D. (1998). The Extended Mind. *Analysis, 58*(1), 7–19.
Di Paolo, E. (2009). Extended Life. *Topoi, 28*, 9–21.
Drayson, Z., & Clark, A. (2019). Cognitive Disability and Embodied, Extended Minds. In A. Cureton & D. T. Wasserman (Eds.), *The Oxford Handbook of Philosophy and Disability*. https://doi.org/10.1093/oxfordhb/9780190622879.013.10.
Kirchhoff, M., & Kiverstein, J. (2018). *Extended Consciousness and Predictive Processing: A Third Wave View*. Routledge.
Levin, J. (2008). Functionalism. In E. N. Zalta (Ed.), *The Stanford Encyclopedia of Philosophy*. Retrieved March 2013, from http://plato.stanford.edu/entries/functionalism
Loughlin, V. (2013). Sketch This: Extended Mind and Consciousness Extension. *Phenomenology and the Cognitive Sciences, 12*(1), 41–50.
Loughlin, V. (2014). *Extended Mind, Extended Conscious Mind, Enactivism*. Doctoral thesis for the University of Antwerp.
Loughlin, V., & Zahidi, K. (2017). What Is Left of the Active Externalism Debate? *European Journal of Philosophy*. https://doi.org/10.1111/ejop.12249
Menary, R. (Ed.). (2010a). *The Extended Mind*. The MIT Press.
Menary, R. (2010b). Cognitive Integration and the Extended Mind. In R. Menary (Ed.), *The Extended Mind* (pp. 227–243). The MIT Press.
Ramsey, W. (2010). Review of Richard Menary (Ed), The Extended Mind. *Notre Dame Philosophical Reviews*. Retrieved March 2013, from http://ndpr.nd.edu/news/24553/?id=22110
Roberts, T., & Columbetti, G. (2015). Extended Affectivity. *Philosophical Studies, 172*, 1243–1263.
Rowlands, M. (2010). *The New Science of the Mind: From Extended Mind to Embodied Phenomenology*. MIT Press.
Rupert, R. (2004). Challenges to the Hypothesis of Extended Cognition. *Journal of Philosophy, 101*, 389–428.
Sprevak, M. (2009). Extended Cognition and Functionalism. *Journal of Philosophy, 106*(9), 503–527.
Sutton, J. (2010). Exograms and Interdisciplinarity: History, the Extended Mind, and the Civilising Process. In R. Menary (Ed.), *The Extended Mind* (pp. 189–226). The MIT Press.
Tribble, E. (2005). Distributing Cognition in the Globe. *Shakespeare Quarterly, 56*, 135–155.
Wheeler, M. (2010). In Defense of Extended Functionalism. In R. Menary (Ed.), *The Extended Mind* (pp. 245–270). The MIT Press.

3

Radical and Sensorimotor Enacted Mind

3.1 Introduction

In Chap. 2, I examined Extended Mind, which is the claim that objects in the environment can, under certain conditions, be part of an individual's cognitive and mental processes. I also showed that what I called First Wave Extended Mind could be read as compatible with the Representational Orthodoxy.

In this chapter, I will examine Enacted Mind. I will focus on two leading enactivist views: Radical Enactive or Embodied Cognition (REC) and Sensorimotor Enactivism. Both views can be read as challenging the Representational Orthodoxy.

3.2 Radical Enactive or Embodied Cognition (REC)

Suppose I ask you to imagine yourself sitting on a sandy beach. I describe to you the cool waves lapping near your feet and the warm summer sun shining above your head. If the Representational Orthodoxy is correct,

V. Loughlin, *4E Cognitive Science and Wittgenstein*, New Directions in Philosophy and Cognitive Science, https://doi.org/10.1007/978-3-030-89463-4_3

then your ability to imagine this scenario is dependent upon the activation of a representation inside your head. The vehicle for this representation is whatever process or set of processes inside your brain that physically implements your imagined thought. The content is what this imagined thought is about, namely that you are sitting on a sandy beach under a warm summer sun. Notice that this vehicle can have properties not possessed by the content and vice versa. For one thing, the vehicle is physically located yet a similar claim cannot be made about the content. For another, the content of your imagined thought is truth evaluable but not so the vehicle. I could, for instance, ask you to describe what you are experiencing as you sit on this beach. If your description involved, say, talk of hurricanes and tornadoes and other natural disasters, then I may conclude that you have not done what I asked you to do and in this sense the content of your imagined thought is false. Yet no such falsity is applicable to whatever vehicle inside your head carries this content.

This representationalist account seems to give us a scientifically credible way to investigate your thought. Suppose I were to open up your head during those moments when you are imagining yourself lying on that beach. I of course will not find your thought. I will only find complex electrochemical interactions occurring inside your brain. Yet if we subscribe to the Orthodoxy, this is not a problem. For the goal here is not the find the thought itself. It is rather to find the process or set of processes inside the brain that act as the vehicle for your imagined thought. And if we have access to the right neuroscientific tools, this seems an entirely achievable aim. Thus, once it is accepted that your imagined thought is dependent upon an internal representation, then your thought can be understood to be a observable item in your brain, which can then be measured and quantified.

However, while vehicles can have properties not possessed by contents and vice versa, vehicles and contents nonetheless come as a package. If so, then it must be possible to give a credible account of how some process or set of processes in the brain can specify, describe, characterize or portray some environmental feature. Much ink has been spilt trying to provide such an account. Yet within the context of 4E, leading proponents of Enacted Mind, such as Daniel Hutto and Erik Myin (henceforth Hutto and Myin), have argued that no such account is available.

Think of the rings of a tree. These rings can be used to gather information not only about the age of the tree but also about the sorts of climate conditions present during the life of the tree. If these rings are far apart, then those conditions were likely warm and wet. If these rings are instead close together, then those conditions were instead dry. There are thus important co-variations between tree rings and climate. And when one state of affairs co-varies with another state of affairs, then this can give us important information about those states of affairs. In the case of trees, we can use these rings to gain information about the severity of droughts in the distant past, for example. Hutto and Myin (2013) label the information gathered in this manner information-as-covariance.

However, information-as-covariance is distinct from information-as-content. This is because information-as-content, according to Hutto and Myin, "has special properties—e.g. truth, reference, implication—that make it logically distinct from, and not reducible to, mere covariance relations between states of affairs. Though covariance is scientifically respectable, it isn't able to do the required work of explaining content" (ibid., p. 67). In which case, content has properties that make it irreducible to information-as-covariance. Hutto and Myin, with a nod to David Chalmers' Hard Problem of Consciousness, term this the Hard Problem of Content.[1] This Hard Problem then poses a serious challenge to representationalist accounts. For while there can be important co-variations between brain processes and bodily behaviours, such co-variations do not and cannot substantiate claims about content.

Instead, information-as-content, insist Hutto and Myin, is scaffolded by our shared social and linguistic practices. If so, then there are no contents outside of these shared practices and given that vehicles and contents come as a package, then there are no such vehicles either. As regards our previous example then, there is no credible way to demonstrate that the content of your imagined thought is carried by some process or sets of processes in your brain, since there are no contents inside your brain. Indeed, Hutto and Myin deny that there are any internal mental contents (Hutto & Myin, 2017, pp. 34–35). They call their proposal Radical Enactive or Embodied Cognition (REC).

Key to this REC proposal is the distinction between basic and non-basic cognition. Basic cognition is non-contentful; non-basic cognition is contentful.

For example, Hutto and Myins' Embodiment Thesis "equates basic cognition with concrete spatio-temporally extended patterns of dynamic interaction between organisms and their environments" (Hutto & Myin, 2013, p. 5). Basic cognition thus is a dynamically unfolding series of interactions between individuals and their environment. Such cognition can be skilful, even sophisticated. Certain forms of remembering, for example, can be understood in terms of basic cognition (Hutto & Myin, 2017).

On the other hand, non-basic cognition involves content. Imagine that you are sitting on a train as you arrive in a foreign city. Before departing from the train, you might start to plan how to get to your hotel. You might think about how you will need to navigate your way through the train station and find where the taxis are. This involves content, since your ability to plan in this way can only be understood by appealing to wider symbol-manipulating practices. For example, it is your participation in such practices that enables you to characterize the taxi rank as likely being close by whenever you exit the train station. If you then exit the station but find that there is no such taxi rank, then your characterization was false or incorrect. Information-as-content is scaffolded by such practices. In which case, what happens within such practices is psychologically discontinuous from what happens outside of such practices.

Nonetheless, this psychological discontinuity is compatible with evolutionary continuity. The evolutionary continuity comes from the phylogenetic story that can be told about the natural origins of information-as-content (Hutto & Satne, 2015). According to RECers, symbol-manipulating practices arise out of the intentionally directed behaviours of individuals as and when they are enacting their environments. Once this phylogenetic story is in place, then an ontogenetic story can be told about how individuals come to be active participants in these symbol-manipulating practices and so can then make claims that, for example, characterize and so are truth evaluable, that is, have content.

The REC proposal thus directly challenges the Representational Orthodoxy. For it rejects the assumption that explaining the human mind always requires invoking internal representations. Indeed, for RECers, like Hutto and Myin, significant portions of intelligent behaviour, like remembering, can be explained in fully non-representational terms.

3.3 Sensorimotor Enactivism

I previously discussed Radical Enactive or Embodied Cognition (REC). However, an alternative reading of Enacted Mind is called Sensorimotor Enactivism. Sensorimotor Enactivism targets perceptual phenomena. Nonetheless, like REC, it does so by challenging the Representational Orthodoxy.

Think of what happens whenever we look at an object. What do we really see? If the object is placed in front of us, then we certainly see the front side of the object. But then we don't see the backside of the object. Still, we continue to see the object as whole. How is this possible? That is, how is it possible to see an object as whole when what we really see is only the front side of the object? This perceptual phenomenon is sometimes called amodal perception (Noë & O'Regan, 2002, p. 578; Noë, 2012, p. 56).

Sensorimotor Enactivists, like O'Regan and Noë, explain amodal perception by arguing that it involves a perceiver's implicit or practical knowledge of the sensorimotor dependencies enacted as and when a perceiver has engaged with or is engaging with an object or objects. These dependencies are sensory in that they refer to the perceiver's various sensory modalities (in this case, vision) and they are motor in that they refer to how these ways of sensing the environment will change as and when the perceiver moves in their environment and/or when an object or objects move in the perceiver's environment. As such, O'Regan and Noë take seriously the detailed ways in which our bodies and objects in our environment don't simply help us perceive but rather are part of what we perceive.

Consider then the visual modality. The sensorimotor dependencies unique to this modality, according to these Sensorimotor Enactivists, will include facts like how our eyeballs have a particular size and shape, which means that whenever our eyeballs saccade, the light hitting off our retinas will alter in ways determined by those saccades (O'Regan & Noë, 2001, p. 941). Or how flow patterns on our retinas will expand whenever we move forwards and contract whenever we move backwards. Or how images on our retina will, under normal circumstances, disappear whenever we close our eyes.

Other dependencies unique to this modality, continue these enactivists, include facts like how objects are always positioned at certain angles from a perceiver's body, which ensures that a perceiver always views an object from a certain distance. Or how, in normal circumstances, perceivers only ever partially see an object. As such, whenever they move around the object, some parts of the object will appear and other parts will disappear. Or how the colour and brightness of the light reflected from the surface of an object will change as the perceiver moves and/or the object moves and/or the light source moves (ibid., pp. 941, 942).

Notice that, on this Sensorimotor Enactive account, it is not the dependencies themselves that are crucial to perception but rather a perceiver's implicit or practical mastery of those dependencies. For example, the fact that our eyeballs saccade is an automatic bodily response and so is not something that can be mastered. However, the way in which those same saccades can alter the light hitting off our retinas is something that can be mastered in an implicit or practical fashion. The Sensorimotor Enactive claim is thus that perceptions are skilful to the extent that such perceptions involve this sort of unhesitating and unthinking implicit or practical mastery or embodied know-how.

Go back then to amodal perception. According to Sensorimotor Enactivists, while processing in the brain is of course needed for the perceiver to see an object as whole despite only seeing the front side of the object, such processing cannot explain how or why the perceiver perceives in this way. Rather, it is the perceiver's implicit or practical mastery of those dependencies enacted as and when they engage with the object, like their mastery of how when they move in relation to the object, some parts of the object will appear while other parts will disappear, that explains how and why they perceive the object as whole.

Contrast this Sensorimotor account with that of a representationalist explanation of amodal perception. Nanay (2018) has argued that amodal perception involves our perceptual system representing the occluded parts of an object. Moreover, this internal representation should be understood in neuroscientific terms, that is, in terms of early cortical processing in the brain. Given that Sensorimotor Enactivism instead explains amodal perception in terms of a perceiver's implicit or practical embodied know-how of those sensorimotor dependencies enacted when a perceiver engages with an object, then it follows that Sensorimotor Enactivism is opposed to such a representationalist account of amodal perception.[2] Sensorimotor Enactivism can thus be read as challenging the Representational Orthodoxy.

However, a leading objection to Sensorimotor Enactivism has come, not from the Orthodoxy itself, but rather from within enactivist circles. Hutto (2005) has argued that Sensorimotor Enactivists face a dilemma. If read one way, the Sensorimotor Enactive account is too strong. If read another way, the Sensorimotor Enactive account is too weak. This is because, as Shapiro (2011) points out, we can distinguish between a strong and a weak reading of implicit or practical knowledge of sensorimotor dependencies. On the strong reading, a perceiver must interact with their environment now in order to perceive: no action, no perception. On the weak reading, it is the perceiver's (perhaps long acquired) skills and abilities to move and interact with an environment that realize their perception. The dilemma then, according to Hutto, is that the strong reading is not plausible whereas the weak reading ensures that Sensorimotor Enactivism is nothing more than a descriptive account (Hutto, 2005, p. 399).

One way to unpack Hutto's dilemma is the following. Note that this is my account of Hutto's objection, not Hutto's account of his objection.

Consider that implicit or practical knowledge of sensorimotor dependencies is an ability. Yet, as Rowlands notes (2010, p. 77), we can distinguish between possession of an ability and exercise of that ability. Given that know-how of sensorimotor dependencies is an ability, then this distinction must apply here. However, on a strong reading, there is no such distinction, since on a strong reading this knowledge must be enacted to be possessed. Compare this to one of Rowlands' own examples (ibid., p. 77). I know how to play piano. But even if I never exercised that ability again, I can still retain that ability. However, on a strong reading my lack of exercise of this ability means that I no longer possess this ability. This simply isn't plausible.

On the other hand, the weak reading is too weak. Think back to amodal perception. Our initial question was: how is it possible for us to see the object as whole whenever we only see the front side of the object? The Sensorimotor answer, on a weak reading, is that the perceiver has acquired the ability to interact with their environment in particular ways. Yet this doesn't explain why I see the object as whole when I only see the front side of the object. If the Sensorimotor reply is simply to point out that a perceiver knows how, if they move around an object, some parts of the object will appear and others will disappear, then this just re-describes the phenomenon. For the phenomenon is the fact that we can only see the front side of the object but not the backside or we can only see the backside but not the front side and yet in both cases we continue to see the object as whole. It seems then that the Sensorimotor Enactive claim, on a weak reading at least, is nothing more than a re-description of amodal perception.

This criticism of Sensorimotor Enactivism suggests that clarifying the role of implicit or practical knowledge of sensorimotor dependencies in perception is far from straightforward. I won't attempt to resolve this issue here. Rather, in Chap. 6 (Sect. 6.3), I will present my own enactivist view (Structural Enacted or Extended Mind or STEEM), which restricts the appeal to such knowledge to the learning of phenomenal concepts. But if so, then we don't need know-how (of any kind) in order to see.[3]

3.4 The Representationalist Objection (RO)

I turn now to a possible way in which proponents of the Orthodoxy could object to the challenges posed by Enacted Mind. I will call this the Representationalist Objection (henceforth RO). If RO is correct, then both REC and Sensorimotor Enactivism could be understood as failing to challenge that Orthodoxy.

RO works by distinguishing between causal relations and constitution and insisting that constitution is a synchronic notion (in a way that I will outline below) whereas causal relations are temporal. Think, for example, of a game of chess. We can distinguish between the history of a particular game, that is, the sequence of moves made by the players of the game, and the current arrangement of pieces on the chessboard. For while the sequence of moves will be part of any explanation for the current arrangement of the pieces on the board, the current arrangement is nonetheless separate from this history. For example, someone may have no interest in how the current arrangement came about but still be very interested in what happens now that the game is set up in a certain fashion.

When it comes to thinking about causation and constitution, so RO continues, we can examine the sorts of complex causal dependencies that may exist between individuals and their environments. These dependencies will be diachronic, that is, extended across time. However, we can also examine those processes that constitute an individual's various cognitive processes. Such constituents will be synchronic, that is, happening at some moment in time. The term synchronic can thus be defined as meaning x only constitutes y when y is spatially and materially coincident with x at time t (I adapt this definition from Wilson, 2007).

If this is accepted, then it poses a problem for enactivists. For enactivist accounts (as both REC and Sensorimotor Enactivism demonstrate) appeal to temporally extended interactions between individuals and their environments. These interactions are extended across time and so are diachronic. These enactivist accounts will thus reveal the causes of human mentality. Yet if constitution is synchronic, then such interactions cannot reveal what constitutes such mentality. Moreover, according to a synchronic view of constitution, the only items that could fit the bill for such constituents are internal representations. For only an internal representation could constitute a cognitive process, in the sense that the cognitive process will be spatially and materially coincident with the internal representation at time t. If so, then these enactivists accounts fail to challenge the Representational Orthodoxy.

Someone who can be regarded as a proponent of RO is Block (2005). Block has challenged Sensorimotor Enactivism on the grounds that even if a perceiver's current perceptions may be caused by implicit or practical embodied know-how of sensorimotor dependencies, this does not show that such know-how also constitutes these perceptions. Indeed, Block insists that Sensorimotor Enactivism fails to resolve what he calls the constitutive question, since no appeal to sensorimotor dependencies can identify the processes that compose an individual's mind.[4]

Recall the example of amodal perception. If Block is correct, then Sensorimotor Enactivists fail to exclude a representationalist account of such perception.[5] For the constituent of the perceiver's perception of the object as whole could be an internal representation, since this perception could be spatially and materially coincident with an internal representation at time t. Nonetheless, the causes of this perception could be diachronic, that is, they could include know-how of the sorts of sensorimotor dependencies enacted as and when the perceiver has interacted with or is interacting with the object.

A proponent of RO could also target REC. Think back to Hutto and Myin's Embodiment Thesis (Sect. 3.2 of the present chapter). According to RO, while the Embodiment Thesis may reveal the causes of basic cognition, it nonetheless fails to reveal what constitutes basic cognition. For if constitution is synchronic, then the constituents of basic cognition cannot be framed in diachronic terms, that is, understood as a dynamically unfolding series of interactions between an individual and their environment. Rather such constituents could be internal representations. In Sect. 3.2, I noted that remembering an environmental feature is, according to Hutto and Myin, an example of basic cognition. If RO is accepted, then such remembering could be spatially and materially coincident with an internal representation at time *t*. But then there is no distinction between basic and non-basic cognition, since basic cognition then involves content. The REC proposal consequently breaks down.

This reveals that if RO is endorsed, then neither Sensorimotor Enactivism nor REC challenge the Representational Orthodoxy. It follows then that if these enactivist challenges are to succeed, then they both need to avoid RO.

However, Hurley (2010, p. 106) has argued that it begs the question against externalist approaches, like Enacted Mind, to assume, without argument, that the causal-constitutive distinction coincides with the boundary between what is internal and what is external. RO arguably makes such an assumption. As such, RO could be accused of begging the question against Enacted Mind. Yet Hurley also insists that the causal-constitutive boundary should be understood in terms of explanation and not in terms of metaphysics (ibid.). Yet it is precisely on this point that Block disagrees. He insists that regardless of explanatory commitments, there is a fact-of-the-matter about what is and what is not part of the minimal metaphysically sufficient condition for perception (Block, 2005). Thus, whereas Hurley takes the issue to be about explanation, Block takes the issue to be about metaphysics. But if so, then it is not clear how this debate can be resolved.

One way then to move this debate forward is to acknowledge that the issue is about metaphysics (and so Block is correct) but nonetheless insist that we can understand that metaphysics in a way that supports rather than challenges enactivist approaches (and so Block is incorrect).[6]

For example, I previously noted that the view of constitution at the heart of RO is synchronic, which I took to mean that *x* only constitutes *y* when *y* is spatially and materially coincident with *x* at time *t*. However, as Gallagher (2017, p. 9) makes clear, constitution, according to enactivists, need not be framed in these terms.[7] This is because constitution can instead be framed as diachronic. A diachronic view of constitution ensures that temporal causal relations can substantiate constitutive claims about those processes that compose the mind. Gallagher puts the point as follows:

> In a dynamical gestalt composed of processes that unfold over time, and characterized by recursive reciprocal causality relations, changes in any processual part (above a certain threshold) will lead to changes in the whole, and changes in the whole will imply changes in the processual parts. In contrast to a synchronic, compositional notion of constitution, these kinds of causal relations are diachronically constitutive of the phenomena. (ibid., p. 10)

If Sensorimotor Enactivists and RECers were to endorse diachronic constitution, then they could avoid RO, given that RO assumes a synchronic view constitution.[8]

For example, it is now straightforward to see how Hutto and Myin could avoid RO. Consider that, according to REC's Embodiment Thesis, basic cognition involves "wide reaching organismic sensorimotor interactions that are contextually embedded. These interactions are assumed to take the form of activity that unfolds across time and which essentially involve individuals engaging with aspects of their environments" (Hutto & Myin, 2013, p. 6). If so, then basic cognition cannot be framed in synchronic terms but rather must be understood diachronically, that is, as extended across time. Once understood diachronically, then the Embodiment Thesis is not vulnerable to RO. For the constituents of basic cognition are the multitude of temporally extended interactions that can occur between an individual and their environment. This then excludes any role for momentary internal representations.

Things are more complicated, however, when it comes to Sensorimotor Enactivism. Still, a possible line of attack against RO can be found in a recent assessment of Sensory Substitution Devices.

Bach-y-Rita's (1972) Tactile Vision Substitution System or TVSS (Noë, 2004; O'Regan, 2011) is what is called a Sensory Substitution Device. This device substitutes visual perception for tactile vibrations on the skin. In more detail, it consists in a head or eyeglass mounted camera, the visual output of which is used to turn on a series of vibrators that are attached to the body of a blind (or blindfolded) individual. The individual is then trained in how to use the TVSS device, that is, they come to learn how bodily movement in an environment with the device changes the patterns of tactile vibrations on their skin. After doing so, individuals begin to report that they can experience objects as arranged in the environment around them. Indeed, Noë (2004, p. 26) points out that such individuals are able to make judgements about the number, relative size and position of objects in the environment. According to Sensorimotor Enactivists then, the extent to which an individual can use this device to master, in an implicit or practical fashion, the dependencies particular to the visual modality will be the extent to which the individual can then "see" using this device.

However, Palermos (2014) has argued that when an individual is using the TVSS device to "see", then the sorts of quasi-visual experiences that can occur as the individual interacts with their environment result, not from the individual's use of the device per se, but rather from the sort of dynamical coupling that emerges between the individual and the device. As such, the individual and the device should not be understood as two separate systems but rather as one non-decomposable dynamical system. Understood in this way, there are then dense reciprocal causal relations between the various parts of this system, such that output from one part of the system becomes input for another part and vice versa.

If Sensorimotor Enactivists, like O'Regan and Noë, were to endorse Palermos' reading of TVSS, then they could reject RO. For the individual and the TVSS device can be understood as forming a dynamical gestalt (to use Gallagher's term). This is because the complex looping now operational within this system ensures that the quasi-visual experiences of the individual concerned are diachronically constituted by densely coupled reciprocal causal relations between the individual and the device that emerge as and when the individual uses the device. If so, then, pace RO, this Sensorimotor Enactive explanation does identify the constituents of these experiences. For the constituents of these experiences will be the temporally extended coupling relations between the individual and the device.[9]

Thus, by clarifying some of the ways in which REC and Sensorimotor Enactivism can be read in terms of diachronic constitution, this then addresses Block's criticism and so avoids RO. If both approaches can avoid RO, then both REC and Sensorimotor can substantiate their respective challenges to the Representational Orthodoxy.

3.5 Conclusion

In this chapter, I have shown that both Radical Enactive or Embodied Cognition (REC) and Sensorimotor Enactivism challenge the Representational Orthodoxy. And while both enactivist approaches potentially encounter a Representationalist Objection (RO), they can, so I have argued, also defend themselves against this objection by appealing to diachronic constitution.

However, if the assessment given in Chaps. 2 and 3 is correct, then Extended Mind and Enacted Mind offer conflicting responses to the Representational Orthodoxy. For whereas Extended Mind can be read as compatible with that Orthodoxy, Enacted Mind can instead be read as challenging that Orthodoxy. And given that both Extended and Enacted Mind are part of the wider 4E family, then it follows that there is a lack of consensus within 4E as to whether or not 4E approaches challenge the Representational Orthodoxy.

The aim then in the coming chapters is to show how, with the help of Wittgensteinian considerations, 4Eers can unify their response to that Orthodoxy and so develop an anti-representationalist account of both Enacted Mind and Extended Mind.

Notes

1. This is not to say that the two Problems are equivalent. For example, Chalmers (1996) has argued that the Hard Problem of Consciousness is not solvable. Hutto and Myin (2013) agree with this assessment. Yet, according to Hutto and Myin, the Hard Problem of Content is only Hard for those who claim that talk of information-as-covariance can substantiate talk about information-as-content. Once this claim is rejected, then the Hard Problem of Content can be resolved. This is because, so Hutto and Myin claim, we can now naturalize content. For example, content can be understood to be a scaffolded feature of shared social and linguistic practices.
2. Others have also attempted to explain Sensorimotor Enactivism in neuroscientific terms. See Seth (2014).
3. See Hutto (2005, p. 401) for an equivalent conclusion.
4. Note here the clear parallel between the coupling-constitution fallacy as detailed in Chap. 2 Sect. 2.4 and Block's objection to Sensorimotor Enactivism. In both cases, it is alleged that no appeal to interaction between an individual and an object or objects in the environment, no matter how complex or extensive, can demonstrate that such interactions are part of those processes that compose the mind.
5. Yet perhaps this requirement is too strong? One could perhaps soften the Sensorimotor Enactivism position and argue that, although it could be compatible with a representationalist account of perception, there are good reasons to deny that it is. But if so, then there is nothing in the letter of Sensorimotor Enactivism that rules out representationalism tout court (thanks to Erik Myin for this suggestion). However, I would be sceptical of this move. This is because I regard Sensorimotor Enactivism's challenge to the Representational Orthodoxy as a central feature of the position. Indeed, I take this feature to be one of the main selling points of the position. Remove that feature and it seems Block is correct.

6. There is a third option. This is to insist that the issue is about identity and so cannot be framed in either causal or constitutive terms. This will be the option pursued by Structural Enacted Mind (see Chap. 6, Sect. 6.3).
7. I reference Gallagher here simply because he states how constitution is understood by enactivists. I make no further claim as to whether or not Gallagher would also endorse a metaphysical response to Block. For my purposes, it will suffice to show that a metaphysical response to Block to available to enactivists and this response involves endorsing diachronic constitution, as Gallagher has articulated that notion.
8. Note that a diachronic form of constitution will not save a Complementarity or Second Wave Extended Mind from the coupling-constitution fallacy. Recall Sutton's reading of Tribble's study of actors in Shakespeare's Globe (see Chap. 2, Sect. 2.4). Think of those moments when these actors were engaged with props in this theatre. There is no suggestion that the properties of the props the actors coupled with changed as and when the actors used those props. A "dynamic gestalt", on the other hand, requires that changes in the parts bring about changes in the whole and vice versa. Yet, on Sutton's reading, it is the very stability or fixed nature of those props that ensures that they can thereby transform the cognition of the actors. As such there is no dynamic gestalt between the actors and the props. This point echoes a criticism of Second Wave voiced by Kirchhoff and Kiverstein (2018). They point out that while transformation is indeed a dynamic process, the transformation Second Wavers appeal to results from the fixed and so non-dynamic properties of environmental objects. Sutton's Complementarity reading exhibits this same feature. If so, then Sutton cannot invoke diachronic constitution to avoid the coupling-constitution fallacy.
9. Palermos (2014) also argues that the cases that matter for Extended Mind are ones that meet what he calls his ongoing feedback loops criterion. In which case, Extended Mind can avoid any coupling-constitution fallacy. I disagree with Palermos' assessment, however. Take TVSS. Even if we allow that the individual and the TVSS device do indeed form a dynamical gestalt, such a gestalt is limited to the individual and the device. It does not also extend to include objects in the individual's environment. In which case, it is not clear what extends in this example. Yet if nothing extends in this example, then it is not an example of Extended Mind, in which case the coupling-constitution fallacy has no clear application.

References

Bach-y-Rita, P. (1972). *Brain Mechanisms in Sensory Substitution.* Academic Press.

Block, N. (2005). Review of Alva Noë, Action in Perception. *Journal of Philosophy, 102*, 259–272.

Chalmers, D. (1996). *The Conscious Mind.* Oxford University Press.

Gallagher, S. (2017). *Enactivist Interventions: Rethinking the Mind.* Oxford University Press.

Hurley, S. (2010). The Varieties of Externalism. In R. Menary (Ed.), *The Extended Mind* (pp. 101–153).

Hutto, D. (2005). Knowing What? Radical Versus Conservative Enactivism. *Phenomenology and Cognitive Sciences, 4*(4), 389–405.

Hutto, D., & Myin, E. (2013). *Radicalising Enactivism: Basic Minds Without Content.* MIT Press.

Hutto, D., & Myin, E. (2017). *Evolving Enactivism: Basic Minds Meets Content.* MIT Press.

Hutto, D., & Satne, G. (2015). The Natural Origins of Content. *Philosophia, 43*, 521–536.

Kirchhoff, M., & Kiverstein, J. (2018). *Extended Consciousness and Predictive Processing: A Third Wave View.* Routledge.

Nanay, B. (2018). The Importance of Amodal Perception in Everyday Perception. *i-Perception, 9*(4).

Noë, A. (2004). *Action in Perception.* The MIT Press.

Noë, A. (2012). *Varieties of Presence.* Harvard University Press.

Noë, A., & O'Regan, K. (2002). On the Brain-Basis of Visual Consciousness: A Sensorimotor Account. In A. Noë & E. Thompson (Eds.), *Vision and Mind: Selected Readings in the Philosophy of Perception* (pp. 567–593). A Bradford Book, MIT Press.

O'Regan, K. (2011). *Why Red Doesn't Sound Like a Bell: Understanding the Feel of Consciousness.* Oxford University Press.

O'Regan, K., & Noë, A. (2001). A Sensorimotor Account of Vision and Visual Consciousness. *Behavioural and Brain Sciences, 24*(5), 939–1031.

Palermos, S. O. (2014). Loops, Constitution and Cognitive Extension. *Cognitive Systems Research, 27*, 25–41.

Rowlands, M. (2010). *The New Science of the Mind: From Extended Mind to Embodied Phenomenology.* MIT Press.

Shapiro, L. (2011). *Embodied Cognition*. Routledge, Taylor and Francis Group.
Seth, A. K. (2014). A Predictive Processing Theory of Sensorimotor Contingencies: Explaining the Puzzle of Perceptual Presence and Its Absence in Synesthesia. *Cognitive Neuroscience, 5*(2), 97–118. https://doi.org/10.1080/17588928.2013.87788
Wilson, R. (2007). A Puzzle About Material Constitution and How to Solve It: Enriching Constitution Views in Metaphysics. *Philosophers' Imprint, 7*(5).

4

Wittgenstein

4.1 Introduction

In PI 185–242, Wittgenstein engages in prolonged and detailed discussion about what it means to follow a rule.[1] The remarks contained within this discussion are now understood to be among the most significant within PI and perhaps within Wittgenstein's overall corpus. Yet determining how to understand these remarks is far from straightforward. Indeed, there is now a vast secondary literature on Wittgenstein's so-called rule following considerations (see Miller and Wright (2002) and Holtzman and Leich (2010) for two edited collections).

I won't attempt to engage with all this literature here (this would require a book of its own). I shall instead focus on one position within that literature, that is, the position defended by Baker and Hacker (2009) in their analytical commentary on PI. In particular, I will focus on Macha's (2015) support and development of Baker and Hacker's position. For my purposes here, I am going to assume that these authors are correct in their assessment of what Wittgenstein says about rule following.

According to Baker and Hacker (2009, p. 87), there is what is called an internal relation between a rule and what accords with the rule, which

V. Loughlin, *4E Cognitive Science and Wittgenstein*, New Directions in Philosophy and Cognitive Science, https://doi.org/10.1007/978-3-030-89463-4_4

is exhibited in the following of a rule. Macha (2015) agrees. He states: "Baker and Hacker are generally right that 'the concept of a rule and the concept of what accords with it (what is a correct application of it) are internally related' and that this view is generally implicit in Wittgenstein's treatment of rules and rule following" (ibid., p. 135). I shall adopt Macha's terminology here. In which case, we can say that there is an internal relation between the concept of a rule and the concept of what accords with it. If so, then following a rule involves enacting an internal relation between these two concepts. As will be explained in this chapter and the next, I use the term "enacting an internal relation" as a philosophical term of art, that is, as a way of focusing on the behavioural criteria needed to be a rule follower.[2]

Baker and Hacker (2009, pp. 50, 152) also demarcate the criteria needed for following a rule from how one learns a rule. This second idea arguably develops the first idea. For if what makes a rule a rule is the internal relation between the concept of the rule and the concept of what accords with it, then the criteria needed to follow a rule will be the enactment of this internal relation and this enactment will be distinct from whatever preceded this enactment, such as how one learned the rule.

Within the context of the rule following literature, these two ideas, along with a host of other considerations, allow Baker and Hacker to reject the sort of communitarian views of rule following famously defended by the likes of Kripke (1982) and Malcolm (1988). For the communitarians, there can be no rule following without the wider agreement of a community of speakers. Baker and Hacker reject this view. They insist that while customs or language-games of course provide the background needed to follow rules, following rules does not depend upon multiple agents arranged in a social group. Rather it depends upon multiple occasions an action is performed (Baker & Hacker, 2009, p. 151; see also Schroeder, 2006). In which case, a solitary individual, like a Robinson Crusoe stranded on a desert island, could potentially follow rules. Think back to the previous two ideas. If the behavioural criteria for following a rule involve the enactment of an internal relation, then a Crusoe could indeed enact this relation on multiple occasions. In doing so, Crusoe would be a rule follower. (As we shall see below, these multiple occasions do not make the rule following a process.)

My aim here is not to adjudicate this dispute between the communitarians and Baker and Hacker. As stated previously, I am going to assume that Baker and Hacker (and Macha) are correct. Rather my aim is to show how the two ideas discussed above—one about internal relations between a rule and what accords with a rule, the other about the distinction between criteria for rules and learning about rules—can be used to reject what I shall call the Process Model. This can be illustrated in the following way.

Take a simple case of learning a rule. A teacher instructs a pupil, maybe via an example or series of examples. Perhaps after some failed attempts, the pupil eventually masters the rule. Such teaching and learning is clearly a sequence of actions spread out in time. So understood, this teaching and learning can be understood to be process. According to the Process Model then, what is true about teaching and learning a rule is also true about the criteria for following a rule. That is, just as teaching and learning a rule is a sequence of actions spread out in time and so a process, then so too is following the rule.

However, if we endorse the two ideas discussed above, then we can reject this Process Model. For once we accept that the criteria for following a rule involves an internal relation between the concept of a rule and the concept of what accords with it, which ensures that the criteria needed to follow a rule is distinct from how one learns a rule, then even if teaching and learning a rule is indeed a process, the same cannot be true of following a rule. This is because internal relations are a temporal whereas processes are temporal. If so, then rule following cannot be modelled in terms of processes. We should thus reject the Process Model. This rejection of the Process Model has significant implications for 4E Cognitive Science (see Chap. 6). I will thus spend this chapter and the next developing it in some detail.[3]

4.2 Internal Relations

The notion of internal relations undergoes significant development over the course of Wittgenstein's career.[4] In the 1930s, Wittgenstein abandons his earlier idea that such relations are metaphysical in nature (see TLP

4.123) and instead begins to regard these relations as grammatical, that is, such relations are between the concepts we employ within our customs or language-games. For example, in his lectures in 1930–1933 (as transcribed by G.E. Moore), Wittgenstein defined an internal relation as that "which holds if the terms are what they are, and which cannot therefore be imagined not to hold" (PO 86). This does not denote some metaphysical relation between two objects. It is instead a consequence of our use of language, that is, within the games we play with language, terms that are internally related are what they are due to the relation between them.

In his Remarks on Frazer's Golden Bough, Wittgenstein gives the following example of such a relation:

> [We] might illustrate an internal relation of a circle to an ellipse by gradually converting an ellipse into a circle; but not in order to assert that a certain ellipse actually, historically, had originated from a circle (evolutionary hypothesis), but only in order to sharpen our eye for a formal connection. (PO 133)

Thus, circles are what they are because they can be converted into ellipses and ellipses are what they are because they can be converted into circles. And this speaks to what you or I mean when we talk about ellipses and circles.

Macha (2015) offers a detailed account of how internal and external relations feature within Wittgenstein's later work. For example, Macha identifies five characteristics of internal relations. Here I will summarize those characteristics and note where Macha's account accords with that of Baker and Hacker.

The first characteristic, claims Macha, is that internal relations hold between concepts, which Macha understands to be complexes. Internal relations themselves cannot be described. Rather they can "only [be] shown in descriptions of these complexes" (ibid., p. 84). Think back to the ellipse and circle. The relation involved here can only be shown through the action of converting an ellipse into a circle or vice versa. Note that in seeing the relation between an ellipse and a circle (or vice versa), one doesn't see the ellipse and see the circle and then see some third item that connects the two. Rather to see the relation between the ellipse and the circle just is to see an ellipse being converted into a circle or vice versa. This can be highlighted by recognizing that, according to Macha, the relata involved in an internal relation "do not need to be restricted to facts; they can also be complex acts, practices, or types of behaviour" (ibid.). Thus, when it comes to the ellipse and the circle, the relata are the actions of drawing an ellipse or drawing a circle.

The second characteristic, says Macha, is that "internal relations are relations that hold in virtue of grammar. Grammatical propositions are either explicit statements of the grammar of a language-game or also—in Wittgenstein's final texts—implicit descriptions of our human form of life" (ibid., p. 83). Baker and Hacker (2009, p. 265) call such statements "norms of description". Given that they are normative, then it is not possible to offer evidence either in favour of or against an internal relation (Glock, 1996, p. 191). Instead, such relations form the framework or structure within which justifying or doubting makes sense. This applies to the ellipse and circle. The formal connection between ellipses and circles is definitive of both ellipses and circles. (This second characteristic relates to the third and fifth characteristics discussed below.)

The third characteristic of internal relations is that they are a temporal as opposed to temporal. Macha (2015, p. 93) refers to this as a criterion of temporality. This is because it provides "a tool or method for determining whether a given sentence expresses an internal or external relation in a given language-game" (ibid., p. 97), since external relations, in contrast to internal relations, will involve "accidental circumstances which are, of course, time-dependent" (ibid., p. 96). Baker and Hacker (2009, p. 279) also regard grammatical statements, that is, statements that show or display internal relations, as a temporal as opposed to temporal. Consider the statement "red is darker than pink". This statement is a temporal, not in the sense of describing some metaphysical property of the colour red, but rather because darker than pink is what makes the concept of red what it is. There thus are no circumstances where red would be other than darker than pink.

The fourth characteristic of internal relations is that they "relate their terms only in virtue of these very terms" (Macha, 2015, p. 102). That is, "the terms of an internal relation are related directly, without any mediation, to each other" (ibid., p. 97). Baker and Hacker (2009, p. 87) also deny that there can be any mediating links between the complexes involved in an internal relation. In terms of the ellipse and circle, the direct, non-mediative nature of the relation between these two complexes means that, as noted previously, one simply has to see this relation.

However, an objection here might be that, say, a child could draw a circle but, for whatever reason, be unaware that this circle could be flattened into an ellipse or vice versa. But if so, then the relation involved here must be external and not internal, since if the relation were internal, then the ellipse and the circle could not be identified independently of each other.

Yet for Wittgenstein, language is always contextual. Indeed, the idea that we can remove a word from its present context of use and it will retain its meaning is one Wittgenstein repeatedly attacked. Internal relations highlight this contextuality. Consider, for instance, how someone might teach a child how to draw a circle. Perhaps the teacher first draws an ellipse and then slowly shrinks the shape until it resembles a circle. In doing so, the teacher is showing or displaying an internal relation between an ellipse and a circle, with the goal of getting the child to see this relation and then enact it via their own (the child's) behaviour. All this means is that the teacher is trying to get the child to see the point of what they (the teacher) are doing. This is of course compatible with there being other contexts when the child simply draws a circle or draws an ellipse.

The fifth and final characteristic, according to Macha, is that internal relations exclude any exceptions. This ties together with the second and third characteristics. For the grammatical nature of internal relations and their exclusion of temporality also requires the exclusion of any exceptions to those relations. Recall the statement "red is darker than pink". There are no circumstances where the concept red could not be darker than pink, since if there were such circumstances, then this would no longer be the concept of red.

This ends my summary of Macha's account of Wittgenstein's notion of internal and external relations. I now turn to some of Wittgenstein's remarks on rule following. I will offer my own reading of these selected remarks. I do so in order to show why we should reject the Process Model discussed earlier.

4.3 Rules and the Process Model

My daughter is of an age now (she is three) that she needs to know how to safely cross the road. One day, when both of us are standing at the traffic lights, I say to her, "you only cross the road at the traffic lights when the little green man is showing". I then show her how to do this. For example, we wait for the cars to come to a stop and for the little green man to appear on the traffic lights. When it appears, we both then cross the road. On a later occasion when standing at the traffic lights, I then ask her, "So how should you cross the road?" If she replies, "you only cross the road at the traffic lights when the little green man is showing" and shows me that she can do this, then she can now follow the rule for safely crossing the road.

Yet Wittgenstein asks, "what if one person reacts to an order and training thus, and another otherwise? Who is right, then?" (PI 206).

For example, what if, despite my previous instruction, my daughter had done something entirely different from what I taught her? Maybe she seems to only follow the rule on some roads but not others. We might then conclude that she has not learnt the rule. However, Wittgenstein warns against this conclusion. This is because my daughter's odd behaviour could in fact be compatible with the rule. For maybe she thinks, "I only cross the road at the traffic lights when the little green man is showing. But this only matters when I am crossing roads near my school. On other roads, I can cross the road wherever I please". Notice that nowhere in my teaching of the rule did I stipulate that the rule is to be applied on all roads. Suppose I now add this stipulation. I tell my daughter, "you only cross the road at the traffic lights when the little green man is showing. And this applies on all roads". This won't work either. For perhaps my daughter now thinks, "I only cross the road at the traffic lights when the little green man is showing and this applies on all roads. However, if I am wearing my yellow dress, then I can cross the road wherever I please".

This is of course problematic. For if further interpretations of my rule can always be found and these interpretations are compatible with my original teaching of the rule, then how can my daughter ever follow the rule? Wittgenstein describes this problem as follows:

> no course of action could be determined by a rule, because every course of action can be brought into accord with the rule. The answer was: if every course of action can be brought into accord with the rule, then it can also be brought into conflict with it. And so there would be neither accord nor conflict here. (PI 201)

In other words, if following the rule means interpreting the rule, then an interpretation can always be found, no matter how outlandish or extravagant, that could be made to fit that rule. But if so, then, as we saw above, this leads to a regress of interpretations. The result is that anything that seems right will now be right, which just means that it no longer makes any sense to talk of being right (or talk of being wrong). Rule following now looks impossible.

However, as Wittgenstein puts it, the mistake is to think that "every action according to a rule is an interpretation" (PI 201). That is, it is a mistake to think that whenever my daughter listens to my teaching, her following the rule depends on her ability to interpret the rule. This is because "every interpretation hangs in the air together with what it interprets, and [as such] cannot give it any support" (PI 198). That is, there are always multiple interpretations of any rule available and these alternative interpretations will always be compatible with the rule. Consequently, if we are to avoid a regress of interpretations, then we need to reject the idea that following rules involves interpreting rules.

Yet consider that I (unlike my daughter) understand perfectly well what the rule means, hence why I say to my daughter, "you only cross the road at the traffic lights when the little green man is showing". But if so, then my thinking about the rule does seem to exclude alternative interpretations of the rule. In which case, while rule following may not depend on my daughter interpreting the rule, it could still depend on my thinking about what the rule means.

This does not help clarify matters, however. For saying that I (the teacher of the rule) have thought about the rule is just to say that I already knew how the rule was to be applied. But this is just another way of restating the rule. It thus cannot help resolve our problem. For if asked what my thinking was, I would have to state it and this statement of mine could still be open to interpretation and so a possible regress.

Nonetheless, we might insist that there must be something in my mind that does exclude these rival interpretations. Suppose this is correct. As Wittgenstein highlights, the idea would then be that "this meaning [of the rule] had in its own way already taken all those steps: that in meaning it, your mind, as it were, flew ahead and took all the steps before you physically arrived at this or that one" (PI 188). But how can anything in the mind predetermine reality in this way, that is, prescribe what must be done? It seems we have just replaced one mystery (how can my daughter follow the rule?) with another (how can something in my mind dictate what my daughter must do?). Rule following has gone from being impossible to being mysterious.

One conclusion we can draw from the discussion so far is this. Consider what happens when my daughter and I stand at the traffic lights. My daughter could interpret the rule I am teaching her. Or I could be thinking about what the rule means. On both views, following a rule refers to the process or processes ongoing as and when we stand together at the traffic lights. Following what was said in the introduction, I call this a Process Model. According to this Model, this process could be my daughter interpreting the rule. Or this process could be my thinking about the rule. Yet, as we have seen, both descriptions are problematic. If my daughter interprets this rule, then this leads to a regress of interpretations. If I appeal to my thinking about the rule, then this replaces one mystery with another. The central difficulty is that once that we introduce a mediator between a rule and applications of that rule, then it is no longer clear what it means to follow that rule (see Baker & Hacker, 2009, p. 87). Yet this is exactly what the Process Model does. This suggests that if we are to clarify what it means to follow a rule, then we need to reject the Process Model. In the next section, I will show why this assessment is correct.

4.4 Enacting Rules

Wittgenstein wrote:

> Let me ask this: what has the expression of a rule—say a signpost—got to do with my actions? What sort of connection obtains here?—Well, this one, for example: I have been trained to react in a particular way to this sign, and now I do react to it. But with this you have pointed out only a causal connection; not explained how it has come about that we now go by the signpost; not what this following-the-sign really consists in. Not so; I have further indicated that a person goes by a signpost only in so far as there is an established usage, a custom. (PI 198)

Wittgenstein's point here is that "there is a way of grasping a rule which is *not* an interpretation", but which, from case to case of application, is exhibited in what we call "following the rule" and "going against it" (PI 201, emphasis in original). This is because "[f]ollowing a rule is analogous to obeying an order. One is trained to do so, and one reacts to an order in a particular way" (PI 206).

Hence, for Wittgenstein, following rules means being trained to react in particular ways. Moreover, this training is only possible because there is an already established way of reacting. Training thus involves becoming part of this custom or language-game. For if "[t]o follow a rule, to make a report, to give an order, to play a game of chess, are customs (usages, institutions)" (ibid.), then following rules means being able to meet the behavioural demands imposed upon you by others.

Think back to our example. I am teaching my daughter how to cross the road. Following Wittgenstein, when teaching her, I am training her to meet certain behavioural demands. For it is established how one should cross the road. This involves waiting at the traffic lights for cars to come to a stop and then only crossing the road when you see the little green man is showing. When my daughter's behaviour matches this established behaviour, then it can be said that she can follow the rule and so can safely cross the road. But if so, then following the rule doesn't require that she interpret the rule. Nor does it require that I think about what the rule means.

However, this can't be the full story. For if following a rule just means matching one's behaviour to the behaviour of others, then one either does this or one does not. Rules then are only predictive. Yet while rules do predict what people will do (see Baker & Hacker, 2009, p. 51), they also do more than this. Consider that I don't simply shrug my shoulders when I see my daughter crossing the road wherever she likes. I instead scold her for potentially endangering her life. Rules then don't just predict what will happen. They also prescribe what should happen. Indeed, an account of rule following devoid of prescription would be no account of rule following at all.

In the introduction to this chapter, I clarified two ideas (which were derived from Baker and Hacker). First, there is an internal relation between a rule and what accords with that rule. Using Macha's terminology, I understood this to mean that there is an internal relation between the concept of a rule and the concept of what accords with it. Second, we can demarcate the criteria needed to follow a rule from how one learns a rule. These ideas are related. For if what makes a rule a rule is the internal relation between the concept of the rule and the concept of what accords with it, then the criteria needed to follow a rule is the enactment of this relation and this enactment will be distinct from how one learned the rule. As I will now show, these two ideas give us an account of the prescriptive nature of rules. Moreover, they show why we should reject the Process Model.

In our example, I say to my daughter, "you only cross the road at the traffic lights when the little green man is showing". We can now give the following description of what happens here. Before doing so, however, note that this description is not intended as a literal account of what happens whenever my daughter learns how to cross the road. It rather serves a philosophical purpose, namely to clarify what it means for her to follow a rule. Thus, talk of "perceiving internal relations" and "enactment of internal relations" should be understood as philosophical terms of art. That is, they are simply meant to focus our attention on the behavioural criteria at work in this example.

Let us allow then that my statement "you only cross the road at the traffic lights when the little green man is showing" is grammatical. My statement is grammatical because it shows or displays an internal relation between two complexes, namely between the statement itself and the action of only crossing the road at the traffic lights when the little green man is showing. The first complex is that of the rule. The second complex is what accords with the rule. My goal then, as the teacher in this example, is to get my daughter to perceive this relation and then enact it via her own verbal and non-verbal behaviour. That is, to grasp the point of what I am teaching her and then show me via her own behaviour that she has understood my point.

This internal relation displays the five characteristics identified by Macha (see Sect. 4.2). First, as stated above, it involves concepts or complexes. Second, the statement "You only cross the road at the traffic lights when the little green man is showing" is grammatical. Third, this statement is a temporal (in the sense identified in Sect. 4.2). Fourth, the relation between the complexes involved here is direct or non-mediative. For example, I show or display this relation via my statement. Fifth, there are no exceptions to my statement. This follows from the grammatical nature of my statement and its lack of temporality.

Suppose then that I have shown or displayed this internal relation via what I have said and my daughter has now perceived this relation. I then ask her, "So how should you cross the road?" If she replies, "you only cross the road at the traffic lights when the little green man is showing" and then shows me that she can do this, then she is showing or displaying the same internal relation. That is, she has clearly got my point and so is now following the rule.

This philosophical rendering of our example then clarifies the prescriptive nature of rules. For in following the rule, my daughter isn't simply matching her behaviour to others. Instead, by enacting an internal relation between the complex of the rule and the complex of what accords with it, she is showing or displaying what counts as correct within the context provided by our custom or language-game (see Baker & Hacker, 2009, p. 155). As such, the prescriptive nature of the rule comes from the internal relations involved. That is, it comes from the grammatical nature of what she says and does in response to my question.

Moreover, by enacting this relation, my daughter is then also sensitive to exceptions to the rule. For example, how there can be circumstances where even though the little green man is showing, she should not cross the road, perhaps because of a speeding car or perhaps because the lights are broke and so they only show the little green man.

In line with an anti-communitarian view of rules, while our customs or language-games provide the background needed to follow rules (if they were no cars, then there would be no roads and so no need to know how to cross the road), we can nonetheless see that it is my daughter's enactment on multiple occasions of the internal relation and not some community wide agreement that makes what she is doing an example of rule following.

What then of the Process Model? As stated previously, I am showing or displaying an internal relation between two complexes, with the aim of getting my daughter to perceive this relation and then enact it via her own verbal and non-verbal behaviour. This is a sequence of actions spread out in time and so can be understood to be a process. However, as stated above, it is only when my daughter enacts the required relation that it can be said that she follows the rule. This enactment is not a process. For internal relations are a temporal whereas processes are temporal. In which case, my daughter's following of the rule, unlike my teaching of the rule and her learning of the rule, is not process. These considerations then support the suggestion identified in Sect. 4.3, namely that we should reject the Process Model.

4.5 Conclusion

In this chapter, I have not addressed the many complex issues that make up the secondary literature on Wittgenstein's rule following considerations. I have instead focused on one position within that literature, namely the position defended by Baker and Hacker, and subsequently developed by Macha. I have then used this position, in particular, the notion of internal relations, to achieve my goal, which was to show why we should reject the Process Model. Put simply, we should not model following rules in terms of processes.

However, as I will show in the next chapter, this Process Model is not confined to following rules but can also be applied to our talk about the mind. Nonetheless, as with the case of following rules, we can use Wittgensteinian considerations to clarify why we should reject that Model.

Notes

1. I shall be using the following abbreviations to refer to works by Wittgenstein: TLP = Tractatus Logico-Philosophicus; PI = Philosophical investigations; PO = Philosophical Occasions; LW II = Last Writings on Philosophical Psychology, volume 2; RFM: Remarks on Foundations of Mathematics. RRP I = Remarks on Philosophical Psychology, volume 1. Further details about these works can be found in the bibliography for this book.
2. Glock (1996, p. 95) makes clear that the relation between concepts and their criteria is internal.
3. Importantly, the rejection of the Process Model should not be equated with rejection of all processes. In Chap. 6 (Sect. 6.4), I argue that the sub-personal processes that underpin the ability to remember can extend to include lines or marks on a page (hence this is an example of Extended Mind), such that disruption of those lines or marks impacts on that ability. In which case, sub-personal processes do play a non-trivial role in the individual's ability to remember. Instead, the Process Model should be rejected because it equates the behavioural criteria needed for mastery of a concept to a process. This is problematic due to the internal relations involved in such criteria. Thanks to Ludger van Dijk for helping me clarify this point.
4. Macha (2015, p. 18) shows that the notion of internal (and external) relations is not unique to Wittgenstein but can be found in the work of Moore, Russell and Bradley. Indeed, Glock (1996, p. 189) states that Wittgenstein acquired the notion from Russell.

References

Baker, G., & Hacker, P. (2009). *Wittgenstein: Rules, Grammar and Necessity: Volume 2 of an Analytical Commentary on the Philosophical Investigations, Essays and Exegesis 185–242* (2nd ed.). Wiley Blackwell.

Glock, H. (1996). *A Wittgenstein Dictionary*. Blackwell Publishers.

Holtzman, S., & Leich, C. (Eds.). (2010). *Wittgenstein: To Follow a Rule*. Taylor & Francis Ltd.
Kripke, S. (1982). *Wittgenstein on Rules and Private Language: An Elementary Exposition*. Harvard University Press.
Macha, J. (2015). *Wittgenstein on Internal and External Relations: Tracing All the Connections*. Bloomsbury Academic.
Malcolm, N. (1988). *Wittgenstein: Nothing Is Hidden*. Blackwell Publishers.
Miller, A., & Wright, C. (Eds.). (2002). *Rule Following and Meaning*. McGill-Queen's University Press.
Schroeder, S. (2006). *Wittgenstein*. Polity Press.
Wittgenstein, L. (1993). *Philosophical Occasions: 1912–1951* (J. Klagge & A. Nordmann, Eds.). Hackett Publishing Company.
Wittgenstein, L. (2001). *Tractatus Logico-Philosophicus*. Routledge.
Wittgenstein, L. (2009). *Philosophical Investigations* (G. M. Anscombe, P. M. S. Hacker, & J. Schulte, Trans.). Wiley-Blackwell.

5

Making Sense of Our Minds

5.1 Introduction

In Chap. 4, I introduced the Process Model. I applied this Model to following rules. Accordingly, following a rule is a process. However, using Baker and Hacker's position on rule following, and Macha's development of their position, I showed why we should reject this Model. In this chapter, I will show that rejecting this Model also allows us to dissolve a problem that our talk of the mind may seem to pose.

5.2 Minds Are Strange Things?

Minds are seemingly strange things. What is the mind? I have one. My wife has one. My three-year-old daughter has one. Even my dog has one. Yet what do I know of these various minds? I know that my dog likes to run in the park. I know that my daughter likes peanut butter sandwiches and splashing in puddles. I also know that my wife, often with a big smile on her face, can point out details in my behaviour that I am entirely

V. Loughlin, *4E Cognitive Science and Wittgenstein*, New Directions in Philosophy and Cognitive Science, https://doi.org/10.1007/978-3-030-89463-4_5

unaware of. What then do I know of my own mind? Sometimes it seems my own mind is as alien to me as anyone else's.

Strange things beget strange questions. If the mind is a thing, then it must have boundaries (an idea familiar to those keen on Extended Mind).[1] Where then are the boundaries of the mind? If I were to open up your head, I would of course find your brain. Brains have obvious boundaries. They begin and end somewhere. But is the same true of the mind? Indeed, what does a boundary of the mind even look like?

Moreover, if the mind is a thing, can this thing be lost? People sometimes say, "I think I am losing my mind". Perhaps because they are suffering from great mental distress. Or perhaps because they have just taken a powerful drug and are now noticing its effects. Or perhaps simply because they are happy and really enjoying the moment. But how exactly can someone lose his or her mind? I sometimes lose my wallet. I obviously place it somewhere but then forget where that somewhere is. Yet it is far from clear how anything similar could happen to my mind. How then can people talk about losing their minds?

Maybe such talk is only metaphorical. People don't actually lose their minds. They merely talk as if they do. However, if a friend were to say to you, with tears in their eyes and their voice breaking, "I think I am losing my mind", you wouldn't take them to be speaking metaphorically. But if so, then maybe we can draw a parallel between our minds and objects. For just as you can lose an object, then so too, it seems, you can lose your mind, since in both cases something you previously thought secure and safe in your possession has now been lost or misplaced. Moreover, in the case of your mind, this loss can have devastating consequences.

What then is this thing we call the mind? Think of your distraught friend. It can seem natural to conclude that the source of their mental distress must have come from something occurring inside their head. That is, whenever they said to you, "I think I am losing my mind", there must have been something happening inside their head that prompted them to make their statement. Maybe, for example, a set of processes firing away in their brain. But if so, then perhaps we can settle one of our previous worries. For we previously wondered what a boundary of the mind would look like. Yet if minds can be identified with brain processes and brain processes have boundaries, then minds also have boundaries.

There thus is nothing mysterious about the boundaries of the mind. The boundaries of the mind just are the boundaries of processes in the brain.[2]

If this assessment is true whenever your friend says, "I think I am losing my mind", then it is likely also true for the many other uses of the word mind. To pick just a few examples, people say things like, "It slipped my mind", meaning that they forgot to do something. Or, "I pushed it to the back of my mind", meaning that they wanted to forget something or at least not think about it. Or, "It's been playing on my mind", meaning that some thought or idea continues to puzzle them. We can take these statements to be the result of various processes occurring inside their brains. In other words, whatever prompts them to make one of these statements is a set of processes currently ongoing inside their heads.

For some, these considerations will be attractive. After all, if they are correct, then our goal is not to investigate some strange thing we call the mind. We also don't need to try and disentangle the ambiguities that may surround the statements we make about our minds. Rather our goal is a practical one, namely to use whatever neuroscientific tools we have to identify brain processes. Moreover, given the sort of credence given to talk of the brain (at least in Western societies), then we may also feel that we have legitimized our task. For gone is highfaluting talk of the mind. All we need to do now is apply some hard science to the brain.

Yet despite this apparent appeal, a problem nonetheless remains. For it is not clear that we can simply dismiss the statements we make about our minds, given that it was these very statements that motivated our investigation in the first place. Moreover, when we examine those statements, it is not obvious that these statements have anything to do with brain processes. When someone says, "I think I am losing my mind", they don't mean they are losing processes in their brain (if this even makes sense). They instead mean that they are suffering great mental distress. Indeed, this distress may be evident in the tears in their eyes, the tone in their voice, the twist of their mouth, and so on. A similar account can be given for the other statements considered earlier. When someone says, "It slipped my mind", they don't mean it slipped their brain processes (again, if this even makes sense). They instead mean that they forgot to do something. And their statement may be accompanied by a gasp of astonishment, a shaking of their head. Finally, when someone says, "I pushed it to the back of my mind" or "It's been playing on my mind", none of these statements concern brain processes. Moreover, the meaning of these statements may be entirely evident in the bodily behaviours of the speakers in question. This should remind us of what is often so characteristic of our talk of our minds, which is that we rarely simply talk about our minds. On most occasions, we instead act out our mental lives via what we say and do.[3]

However, this seems to leave us with a problem. On the one hand, something inside us prompts us to make our various statements about our minds. In which case, we need to examine processes occurring in our brains. Yet, on the other hand, when we look closely at our talk about our minds, we find that such talk is not about brains processes at all. Moreover, if we restrict our focus just to brains, then we risk losing sight of what is so distinctive about such talk, namely that we don't talk about our minds but rather we act out our mental lives. But if so, then talk about our minds seems to both direct us inwards towards the brain and outwards towards our bodily behaviours. How then are we to make sense of our talk about the mind?

Now, one possible response to this problem is to insist that talk about our minds is so much "folk psychology", which can and should be replaced once we know more about the brain. I will adopt a very different strategy. Using Wittgensteinian considerations, I will show how we can dissolve the aforementioned problem. Before doing so, however, I will show how the Process Model identified in Chap. 4 can be applied to mental phenomena, for example, it can be applied to pain experiences.

5.3 Feeling Pain

According to Wittgenstein, "it's characteristic of our language that the foundation on which it grows consists in steady ways of living, regular ways of acting. The function is determined above all by action, which it accompanies" (PO 397). Or more simply perhaps, "Language—I want to say—is a refinement. 'In the beginning was the deed'" (ibid., p. 395). Picking up on this last remark, this doesn't entail that what is done prior to language is meaningless. According to Wittgenstein, it can indeed be meaningful (as we will see below). Nonetheless, there is an intimate link between action and language. As Wittgenstein puts it, "The basic form of the game must be one in which we act" (ibid., p. 397).

Recall what was said before about following rules (see Chap. 4, Sects. 4.3 and 4.4). In the example discussed in that chapter, I am teaching my daughter how to cross the road. According to our philosophical rendering of this example, following the rule involves showing or displaying an internal relation. My goal then as the teacher in the example is to get my daughter to perceive this relation. She then follows the rule whenever she enacts this internal relation via her own verbal and non-verbal behaviour. Compare this description to the following remark by Wittgenstein:

> [The learner] does not say: I realized that this happens. Rather: [the learner says] that it must be like that. This "must" shows what kind of lesson he has drawn from the scene. The "must" shows that he has gone in a circle…This must shows that he has adopted a concept…He has read off from the process, not a proposition of natural science but, instead of that, the determination of a concept. (RFM VI, 7; RFM, VI, 8)

This remark accords with the distinction between the criteria for following a rule and how one learns a rule (see Chap. 4, Sects. 4.1 and 4.4). We can put this as follows. How one learns a rule involves completing the learning circle. This is a process. By contrast, the criterion for following a rule involves the determination of a concept. This is not a process. In the context of our example, my daughter follows the rule when she enacts the internal relation. Internal relations are a temporal and so cannot be understood as processes, since processes are temporal.

Yet this rejection of processes is not restricted to following rules. It also applies to some pain behaviours. For the criteria for such behaviours also involve the mastery of a concept, that is, they involve the enactment of an internal relation. I will demonstrate this by developing our previous example. As before (see Chap. 4, Sect. 4.4), my development of this example is not intended as a literal description of my daughter's behaviour. It rather serves a philosophical purpose, namely to clarify the behavioural criteria at work in this example.

As every parent knows, young children fall over all the time. They rarely look where they are going. They have no awareness of the dangers that, to a parent's eyes at least, seemingly lurk around every corner. My three-year-old daughter is no exception. Cutting her hands, bruising her knees, skinning her elbows—these are all regular occurrences in our household. When she falls over and hurts herself, she will often turn to my wife or I for comfort. In most cases, this comforting usually results in our daughter quickly forgetting about her pain and then continuing to play. To my mind, this raises an interesting question. For what happens at that moment? Does, for example, the comforting my wife and I provide dissipate our daughter's pain or somehow make it go away?

One way to approach this question is to consider another one, namely how do my wife and I teach our daughter the use of the word pain? For example, when my daughter falls over and then turns to us for comfort, we may comfort her by saying things like, "foot sore" while pointing at her foot. Our daughter may then point at her own foot and repeat the words "foot sore". We thus seem to be teaching our daughter how to describe her own pain experience. That is, our daughter has a pain experience and what we are doing is enabling her to give a name to this feeling. For if she can name it, then perhaps she can distance herself from it and so return to her playing.

However, Wittgenstein denies that this is the case. He wrote:

> How do words refer to sensations?…Here is one possibility: words are connected with the primitive, natural, expressions of sensation and used in their place. A child has hurt himself and he cries; then adults talk to him and teach him exclamations and, later, sentences. They teach the child new pain-behaviour. "So you are saying that the word 'pain' really means crying?"—On the contrary: the verbal expression of pain replaces crying, it does not describe it. (PI 244)

I think there are two key ideas in this remark. First, when it comes to pain, we have natural pain behaviours, such as wincing, grimacing, crying out or clutching an injured body part. These behaviours are bodily and meaningful. Yet, second, there can be circumstances where words can be "used in their place", that is, "the verbal expression of pain replaces crying, it does not describe it". Thus, when our daughter learns how to say the phrase "foot sore" while pointing at her own foot, she is not describing her pain experience. Rather her statement is itself a verbal expression of her pain experience. This is new pain behaviour. Wittgenstein is thus asking us to "break with the idea that language always functions in one way, always serves the same purpose: to convey thoughts—which may be about houses, pains, good and evil, or whatever" (PI 304).[4]

I previously discussed Macha's (2015) account of internal relations (see Chap. 4, Sect. 4.2). Internal relations, according to Macha, have five characteristics. They relate concepts, that is, complexes, which can be facts, actions, practices or types of behaviour. When statements are grammatical, they can exhibit internal relations between complexes. Moreover, such relations are a temporal, direct or non-mediative and admit of no exceptions. If we apply Macha's account of internal relations to Wittgenstein's remark about new pain behaviour, then this reveals that being taught new pain behaviour is not about describing or naming some pain experience. It is instead about mastering a concept.

Go back then to our example. Our daughter falls over and hurts herself. She then turns to my wife or I for comfort. We provide this comfort, which leads our daughter to carry on playing. Does our comforting dissipate our daughter's pain or somehow make it go away? This was our question.

Notice that a Process Model has obvious application here. According to this Model, what makes an experience one of pain (and not something else) is the process actualized during those moments one has that experience. This seems true of our example. For the question raised in the previous paragraph only makes sense on the assumption that our daughter has some form of experience during those moments when she falls over and our comforting then impacts on that experience. This experience can then be understood to be a process.

However, using the previous Wittgensteinian considerations, we can give a very different account of this example. First, our daughter seeks comfort when she falls over because seeking comfort is part of her natural pain behaviour. Moreover, as her parents, providing that comfort is part of our natural pain behaviours. These natural behaviours are meaningful (as any parent will attest). Nonetheless, second, my wife or I can teach our daughter new pain behaviour. We may thus say, "foot sore" while pointing at our daughter's foot. When our daughter repeats this statement and action back to us, then her statement is grammatical, that is, it shows or displays an internal relation between two complexes, namely between the statement "foot sore" and the action of pointing at our daughter's foot. She has now mastered a concept, which is to say that she now has the ability to verbally express her pain experience.

If this is correct, then we should reject the Process Model. For while our daughter will cry, wince, grimace, and so on when she falls over, it is only by showing or displaying the complexes described above that it can be said she has mastered the concept, since this is the behavioural criterion for the concept. But if so, then her new pain behaviour cannot be modelled in terms of processes, since internal relations are a temporal and so cannot be understood as processes. Once we reject this Model, then this allows us to resolve our previous question. For the comfort we provide neither dissipates our daughter's pain nor makes it go away. Rather our daughter seeks our comfort and we provide it because these are part of our natural pain behaviours. Nonetheless, something new happens whenever we teach our daughter new pain behaviour. In which case, not all of our daughter's pain behaviours are the same. Some are bodily and have no criteria. Others are expressed through language and so do have criteria.

In Chap. 6 (Sect. 6.3), these considerations will play a significant role in my discussion of Structural Enacted Mind. For now, I will return to our problem about the mind and show how that problem can be dissolved.

5.4 Making Sense of Our Talk About the Mind

In Sect. 5.2 of this chapter, I outlined a possible problem concerning our talk about our minds. When we examine our talk about our minds, it seems we need to look inward at what is going on inside the brain. Yet examining such brain processes risks losing sight of the fact that we don't simply talk about our minds. Rather we act out our mental lives via what we say and do. How then are we to make sense of our talk about our minds?

In order to address this problem, I am going to offer reminders about those circumstances when we do talk about our minds. Recall the sorts of examples listed in Sect. 5.2. We say things like, "I think I am losing my mind", "It slipped my mind", "I pushed it to the back of my mind" or "It's been playing on my mind". What then are circumstances when it would make sense to say these various things?

Suppose, for example, I say to my wife, "I think I am losing my mind". She will (hopefully!) ask me why I feel the way I do. I might reply by saying, "My job is very stressful" or "I am worried about tomorrow's appointment with the doctor". These latter statements arguably justify my initial statement. I claim that this is because these statements are grammatical. For they show or display an internal relation between these statements and my initial statement, that is, there is an internal relation between my statement, "I think I am losing my mind" and the statement "My job is too stressful" or the statement "I am worried about tomorrow's appointment with the doctor".

Yet perhaps my wife does not ask me why I feel the way I do, not because she doesn't care, but simply because it is entirely evident from how I express myself what I am actually feeling. Perhaps my voice shakes as I speak. Or perhaps I have a slight tremor in my hands and so I rub them together in an anxious manner. Or perhaps I am hunched over as I speak and seemingly refuse to look her in the eye. Going further, maybe when I make my statement, I seem to both whisper it, as if it were a shameful realization, and shout it out, as if exhausted by the effort. Any of these non-verbal and verbal behaviours could be taken as evidence for how I am feeling, in which case no follow-up statements are needed.

Still, even under these circumstances, I am showing or displaying various internal relations. For what is it about these behaviours that convinces my wife I am being genuine? It is arguably because my wife can perceive the internal relations that I am showing or displaying. For example, she can perceive the internal relation between my statement, "I think I am losing my mind" and my bodily behaviour of partially whispering and partially shouting the words of that statement. As with the example discussed in Chap. 4 (see Sects. 4.1 and 4.4), the term "perceive the internal relations" is a philosophical term of art meant to highlight the behavioural criteria at work in this example.

I could of course be deceiving my wife. Perhaps I am not really in any danger of losing my mind but I merely want to convince her that this is the case, maybe because I am feeling especially vulnerable and so want some sympathy. Nonetheless, even here, internal relations still remain paramount.

Consider that, for Wittgenstein, "'That seems genuine' only makes sense if there is a 'That is genuine'" (LW II 86e). In other words, we can only conceive of me deceiving my wife because there are other occasions when I am not deceiving my wife. The previous paragraphs detailed what these latter occasions would involve, namely me showing or displaying an internal relation between various complexes, such as internal relations between statements or internal relations between statements and bodily behaviours. My wife then perceives these internal relations. If so, then internal relations are what ensure that there can be situations where I can potentially deceive my wife. Consider that if there were no such structures in place, that is, there were no occasions when there is no doubt for my wife that I was being genuine, then all talk of me failing to be genuine and so of attempting to deceive my wife would cease to make sense.

I claim that internal relations also play a key role in the other examples listed earlier. Take one of those examples. Perhaps my wife says to me, "I pushed it to the back of my mind". As noted in Sect. 5.2, this is usually taken to mean that the speaker wants to forget about something or at least not think about it. Suppose I then ask my wife a follow-up question, namely, "What was it that you wanted to forget about?" At this point, my wife may say various things: "I have a meeting with my boss" or "The deadline for the project is next week". Following what has been said above, my wife is here showing or displaying an internal relation between these various statements. It is thus my perception of this relation that allows her latter statements to justify her initial statement. This is the behavioural criterion at work here.

Moreover, if my wife makes her statement with a dismissive wave of her hand or by turning away her head or by grimacing, then I may be in no doubt about her genuineness, since I can perceive the internal relation between her statement and these bodily gestures. Of course, turning the tables, my wife may be trying to deceive me here. Still, if there were no occasions when I had no doubt about her genuineness, then all talk of her failing to be genuine would cease to make sense.

Summarizing so far, we can thus say that what gives sense to our statements about our minds are actions that involve internal relations, that is, internal relations between statements or internal relations between statements and bodily behaviours. In which case, it is the fulfilment of these behavioural criteria that gives these statements and their accompanying behaviours their meaning.

Yet an objection to this summary might be the following. Suppose that when I said to my wife, "I think I am losing my mind", I was sitting inside a brain scanner. Suppose further that a skilled neuroscientist was examining my brain activity when I made my statement. Suppose further still that after careful consideration of my brain activity, the neuroscientist identifies a particular set of processes occurring in some region of my brain when I make my statement. Or suppose I am not sitting inside any brain scanner at all. Rather a psychologist is examining my bodily behaviour when I make my statement. The psychologist might note that, as pointed out earlier, I both seem to whisper my statement, as if it were a shameful realization, and shout it out, as if exhausted by the effort.

The neuroscientist or the psychologist could then claim that it is these brain processes and/or bodily behaviours that make my statement a statement about my mind. Their argument may be that without these processes and/or behaviours, I would not have expressed myself in the way that I did. Moreover, the psychologist might insist that their findings confirm that we act out our mental lives. For I am indeed acting out these behaviours when I make my statement. But if so, then it is a mistake to appeal to internal relations in order to make sense of our statements about our minds. All we need to recognize are the brain processes and/or bodily behaviours involved.[5]

Two things can be said in response to this objection. First, I am not denying that brain processes and/or bodily behaviours play key roles in our mental lives (see Chap. 6, Sect. 6.5 where I discuss the brain). Damage to our brains and bodies have obvious deleterious effects on our abilities to think and feel. However, second, the neuroscientist and psychologist are not simply saying that brains and bodies matter when it comes to what we say about the mind. They are instead saying that without certain brain processes and/or behaviours, I would not have expressed myself in the way that I did. This is a sufficiency claim. That is, their claim is that the activation of certain brain processes and/or presence of certain behaviours were sufficient for me to say, "I think I am losing my mind".[6]

However, while brain processes are certainly necessary for me to make my statement, the sufficiency claim about such processes and/or behaviours can be disputed. Recall Macha's criterion of temporality (see Chap. 4, Sect. 4.3). This criterion allows us to distinguish between internal or external relations. This is because internal relations are a temporal whereas external relations are temporal.

Return then to the neuroscientist and the psychologist. Any brain processes the neuroscientist identifies or any bodily behaviours the psychologist identifies will only be externally related to my statement "I think I am losing my mind". This is because these brain processes and bodily behaviours are temporal, that is, they occur during those moments when I make my statement. If so, then the relation between these processes and behaviours and my statement will be non-direct and so meditative, which is to say that the relata of this relation can occur entirely independently of one another. What this means is that the brain processes the neuroscientist identifies could have occurred but I may not have made my statement. Or the bodily behaviours the psychologist identifies could have occurred but I may not have made my statement. Note that this separation between these brain processes and bodily behaviours and my statement is distinct from what is involved in an internal relation. In an internal relation, the relata are what they are due to the internal relation and so cannot be identified independently of each other (see Chap. 4, Sect. 4.2).

If so, then all the neuroscientist and psychologist have demonstrated is that there is a co-variation between those brain processes or those bodily behaviours and my statement. These co-variations can certainly be of scientific interest. Yet they do not confirm the sufficiency claim, given that, as noted above, such processes and behaviours can occur entirely independently of my statement. This failure to confirm the sufficiency claim means that the objection cited by the neuroscientist and the psychologist can be disputed.

How then are we to make sense of our talk about our minds? As stated previously, what gives sense to our statements about our minds are actions that involve internal relations, which is to say that it is the fulfilment of behavioural criteria that gives these statements their meaning. But if so, then this allows us to dissolve our initial problem. For talk about our minds only seems to pose a problem because we assume that something must happen when we make our various statements about our minds. For example, what happens must be a set of processes occurring inside the brain. This assumption, as should be clear by now, is an example of the Process Model. However, if what gives sense to our statements about our minds are actions that involve internal relations and internal relations are a temporal, then these statements cannot be understood in terms of processes, since processes are temporal. We can thus reject the Process Model. Once we reject this Model, then there is no problem when it comes to our talk about our minds. Our statements about our minds have meaning because of the contexts within which we make those statements.

5.5 Conclusion

In this chapter, I identified a problem that our talk about our minds seems to pose. I then showed how we could dissolve that problem once we rejected the Process Model.

In the next and final chapter, I will show that the Process Model, as identified in Chaps. 4 and 5, also applies to the debates surrounding Enacted Mind and Extended Mind. Using then the sorts of Wittgensteinian considerations discussed in Chaps. 4 and 5, I will argue that proponents of both Enacted Mind and Extended Mind should reject this Model. If they do so, then they can potentially unify their respective responses to the Representational Orthodoxy.

Notes

1. The qualification "If" plays an important role here and should be taken to indicate that I do not support the idea that the mind is a thing. Indeed, as I show in Sect. 5.4 of this chapter, the idea that mind is a thing, along with the problem it seemingly gives rise to, is ultimately in need of dissolution.
2. This sort of move has obvious links with the Representational Orthodoxy. See Chap. 3, the beginning of Sect. 3.2 for such a link.
3. This is the familiar Wittgensteinian point about the inherent contextuality of language (see Chap. 4, Sect. 4.2). For example, the differences between the statements "I think I am losing my mind" and "I think I am lost" or between "It slipped my mind" and "I slipped on a banana" or between "I pushed it to the back of my mind" and "I pushed it to the back of the bookshelf" are all made apparent by the different contexts in which we use these statements.
4. Wittgenstein is of course not providing here anything like a hypothesis about how children do in fact learn the word pain. On the contrary, his talk of possibility should remind us that his aim here (as with his remarks generally) is philosophical, that is, to clarify how a concept like pain may work.
5. Consider the following quote from the neurosurgeon Dr Henry Marsh: "I often have to cut into the brain and it is something I hate doing. I cut into it with a small scalpel and make a hole through which I push with a fine sucker. The idea that my sucker is moving through thought itself, through emotion and reason, that memories, dreams, reflections should consist of jelly, is too strange to understand" (2014). I take Dr Marsh's identification of our minds with our brains to be part of what has been described elsewhere as a "neuroculture" wherein "[n]otions of the brain as a metonym for identity and self are becoming widespread" (Frazzetto & Anker, 2009, p. 819). As part of this culture, we are "cerebral subjects" (Vidal, 2009). Personhood is thus a matter of brainhood. We don't have brains. We are our brains. The neuroscientist discussed in my text can thus be regarded as someone immersed in this wider trend.
6. For a discussion of the difference between necessary and sufficiency, see Crane (2003, p. 14).

References

Crane, T. (2003). *The Mechanical Mind: A Philosophical Introduction to Minds, Machines and Representation* (2nd ed.). Routledge.

Frazzetto, G., & Anker, S. (2009). Neuroculture. *Science and Society, 10*, 815–821.

Macha, J. (2015). *Wittgenstein on Internal and External Relations: Tracing All the Connections*. Bloomsbury Academic.

Marsh, H. (2014). *Do No Harm: Stories of Life, Death and Brain Surgery*. Orion Publishing Group.

Vidal, F. (2009). Brainhood, Anthropological Figure of Modernity. *History of the Human Sciences, 22*(1), 5–36.

Wittgenstein, L. (1983). *Remarks on the Foundations of Mathematics* (G. H. von Wright, R. Rhess, & G. E. M. Anscombe, Eds.). MIT Press.

Wittgenstein, L. (1992). *Last Writings on the Philosophy of Psychology: The Inner and the Outer* (Vol. 2). Blackwell Publishing.

Wittgenstein, L. (1993). *Philosophical Occasions: 1912–1951* (J. Klagge & A. Nordmann, Eds.). Hackett Publishing Company.

Wittgenstein, L. (2009). *Philosophical Investigations* (G. M. Anscombe, P. M. S. Hacker, & J. Schulte, Trans.). Wiley-Blackwell.

6

Structural Enacted or Extended Mind (STEEM)

6.1 Introduction

In Chap. 3, I showed that Extended Mind and Enacted Mind offer conflicting responses to the Representational Orthodoxy, that is, conflicting responses to the assumption that explaining the human mind always requires appealing to internal representations.

For example, First Wave Extended Mind is the claim that cognitive processes extend outside the head when there are conditions of parity or equivalence between what happens inside the head and what happens outside in the environment. As such, First Wave Extended Mind is compatible with the Orthodoxy, since First Wave can be read as the claim that the boundaries of internal representations can, under conditions of parity or equivalence, extend to include usage of objects in the environment.

By contrast, Enacted Mind challenges the Orthodoxy. Radical Enactive or Embodied Cognition (REC) rejects the assumption that explaining the human mind always requires invoking internal representations. Sensorimotor Enactivism rejects representationalist explanations of perceptual phenomena. This is because, according to Sensorimotor Enactivists, perception involves a perceiver's implicit or practical

V. Loughlin, *4E Cognitive Science and Wittgenstein*, New Directions in Philosophy and Cognitive Science, https://doi.org/10.1007/978-3-030-89463-4_6

knowledge of the sensorimotor dependencies enacted as and when a perceiver has engaged with or is engaging with an object or objects in the environment. In which case, perceiving an object or objects does not require that the perceiver represent that object or those objects inside their head.

Thus, whereas First Wave Extended Mind is compatible with the Representational Orthodoxy, both REC and Sensorimotor Enactivism challenge that Orthodoxy. Given that both Extended and Enacted Mind are part of the wider 4E family, then it follows that there is a lack of consensus within 4E as to whether or not 4E challenges the Representational Orthodoxy. Some think it does. Others think it does not.

However, I ended Chaps. 3 and 5 with the promise that, with the help of Wittgensteinian considerations, 4Eers can unify their responses to that Orthodoxy. The aim in this chapter then is to fulfil this promise. In order to do this, I will first identify the conditions that both Enacted Mind and Extended Mind need to meet if they are to challenge the Orthodoxy. I will then present a theoretical framework that, so I claim, meets these conditions.

6.2 Challenging the Representational Orthodoxy

As we saw in Chap. 3, Enacted Mind looks potentially vulnerable to a Representationalist Objection (RO). RO distinguishes between causal relations and constitution and insists that causal relations are temporal whereas constitution is synchronic. A proponent of RO can then use this distinction to argue that since both Sensorimotor Enactivism and REC appeal to, among other things, temporally extended interactions between individuals and their environments, then while these enactivist accounts may reveal the causes of human mentality, they nonetheless fail to reveal what constitutes such mentality. Moreover, the only items that fit the bill for such constituents are, according to RO, internal representations. In which case, neither Sensorimotor Enactivism nor REC successfully challenges the Orthodoxy.

I then showed how these enactivists could avoid RO. This required endorsing a diachronic (as opposed to synchronic) view of constitution. According to this diachronic view, densely coupled reciprocal causal relations constitute the processes that compose the mind. If so, then causal relations can be constitutive. Both REC and Sensorimotor Enactivism can be read as endorsing this view. If so, then both versions of Enacted Mind can avoid RO and so can maintain their respective challenges to the Orthodoxy.

However, diachronic constitution is committed, if only tacitly or implicitly, to a Process Model. As Chaps. 4 and 5 made clear, the Process Model works by claiming that there is some process that is actualized when, for example, one follows a rule, has a pain experience or makes a statement about the mind. This process is then understood to be what makes a rule a rule, a pain experience one of pain or a statement about the mind mental. Diachronic constitution, as we have seen, appeals to processes extended across time (see Chap. 3, Sect. 3.4). But if so, then diachronic constitution is committed in some fashion to the Process Model. Indeed, it is not clear what it would mean to reject the Process Model but still hold on to diachronic constitution (as we will see in Sect. 6.4 of the present chapter).

Using Wittgensteinian considerations, Chaps. 4 and 5 offered reasons as to why we should reject the Process Model. Yet if diachronic constitution and the Process Model are linked, then it follows that we should also reject diachronic constitution. However, once diachronic constitution is rejected, then it seems Enacted Mind is once again vulnerable to RO.

It follows then that if Enacted Mind is to challenge the Representational Orthodoxy, then it needs to be read in a way that fulfils two conditions. First, it must avoid RO. Second, it must do so without endorsing diachronic constitution. An equivalent analysis can also be applied to Extended Mind.

In what follows, I will outline a theoretical framework that, so I claim, fulfils both conditions. This framework rejects attempts to picture the mind in terms of either causal relations and/or constitution. Rather, according to this framework, phenomenal experiences and other mental episodes can be identified with actions that involve internal relations. Moreover, the sub-personal processes that underpin these experiences and episodes can be understood as sometimes extending to include our use of objects. This framework thus brings together elements of both Enacted and Extended Mind in a way that avoids RO without endorsing diachronic constitution. It thus provides a way to unite Enacted and Extended Mind against the Representational Orthodoxy. I call this framework Structural Enacted or Extended Mind or STEEM.[1]

6.3 Structural Enacted Mind

According to Chalmers' (1996) Hard Problem of Consciousness, it is not clear how any explanation of physical processes can explain the what-it-is-like (Nagel, 1974) aspects of phenomenal experience. This suggests that there is an alleged explanatory gap (Levine, 1983) between physicalist accounts of behaviours, on the one hand, and phenomenal experiences, on the other.

Sensorimotor Enactivists deny that there is any such problem. This gap only emerges, so they claim, for those who insist that phenomenal experiences are generated by brain activity (recall their challenge to the Representational Orthodoxy outlined in Chap. 3, Sect. 3.3). If we instead go outside the head and examine the complex ways in which we interact with our environments, then there is no such gap and so no problem. As O'Regan puts it, there is in fact a "natural relation" (O'Regan, 2009) between how we engage with our environments and what we experience.

For example, O'Regan (2009) writes: "to every aspect of phenomenology…there corresponds an objective physical characteristic of the interaction" (ibid., p. 588). Take the example of squeezing a sponge. The softness I feel as I squeeze the sponge, insists O'Regan (2011), just is to interact with an object like a sponge in a certain manner.[2] Thus, the natural relation between action and phenomenal experience is a relation of identity.

Proponents of Radical Enactive or Embodied Mind (REC) adopt a similar position. Hutto and Myin (2013) have argued that "the phenomenal character of experiences must, ultimately, be understood by appealing to interactions between experiencers and aspects of their environment" (ibid., p. 176). This is because there are not two distinct relata—the experiential, on the one side, and the physical, on the other (ibid., p. 169). Rather experiences can be identified with what agents say and do. Moreover, these identities cannot and do not need to be explained. This is because "[n]aturally occurring 'what it is like' or 'what it feels like' illocutions take activities as their natural objects. When we describe phenomenal experience, we cannot help but mention environment-involving interactions" (ibid., p. 177). There thus is no explanatory gap in need of closing. In this sense, both Sensorimotor Enactivism and REC offer equivalent approaches to gap worries.[3,4]

Yet while these enactivist approaches arguably avoid any explanatory gap, they nonetheless raise an issue of their own (one which, e.g., Hutto and Myin do not consider). Consider that no enactivist claims that all actions are phenomenal. But if so, then it follows that there are actions that are not phenomenal, on the one hand, and identities between experience and action, that is, actions that are phenomenal, on the other. As previously discussed, once these identities are in place, then there is no gap between action and experience. Moreover, these identities cannot and do not need to be explained. But even if this is accepted, it still leaves a further question, namely where do these proposed identities between action and experience come from? Until this question is appropriately dealt with, it is not clear how viable it is to claim that there are identities between action and phenomenal experience.

I propose to answer this question using Structural Enacted Mind.[5] According to Structural Enacted Mind, identities between action and phenomenal experience come from the fulfilment of behavioural criteria. This is because the behavioural criteria needed for mastery of phenomenal concepts involves enacting internal relations. These relations are definitional of phenomenal experiences. We can thus identify the enactment of an internal relation with a phenomenal experience. If so, then experiences are structural. They are what they are due to the internal relations involved (the same point can be applied to a mental episode like remembering—see Sect. 6.4 of the present chapter). I will illustrate this using the following two examples. As with the examples discussed in Chaps. 4 and 5, the purpose of these examples is philosophical, namely to clarify the behavioural criteria at work in these examples.

Suppose I were to say, “This sponge is soft” while simultaneously squeezing a sponge. Here my statement can be understood to be a description of what I am feeling as and when I squeeze the sponge. Recall Macha’s criterion of temporality (see Chap. 4, Sect. 4.2). Internal relations are a temporal whereas external relations are temporal. In terms of our example, the relation between my statement and my action is external, since this relation only occurs during those moments when I am squeezing the sponge. Moreover, since this relation is external, then the relata involved in this relation can occur entirely independently of one another (recall the response to the neuroscientist and psychologist discussed in Chap. 5, Sect. 5.4). For example, it is possible that I could squeeze the sponge but not say, “This sponge is soft”, perhaps because the sponge is in fact partially frozen.

However, suppose I am instead asked what softness feels like. By way of reply, I get the questioner to squeeze a sponge while I say, “That is what softness feels like”. Under these circumstances, my statement is not contingent. I am instead defining what softness feels like. As such, my statement is grammatical. Given that my statement is grammatical, then my statement shows or displays an internal relation between two complexes, namely an internal relation between the statement itself and the action (performed by the questioner) of squeezing the sponge. The experience of softness is thus identified with an action, that is, the action of enacting this internal relation. In which case, the identity between action and experience comes from the fulfilment of this behavioural criterion.

This second example is clearly distinct from the first. In this second example, the questioner simply has to see the internal relation being shown or displayed. Consider that if the questioner were to insist that they still didn’t know what softness felt like, then no further demonstrations by me (or anyone else) could help them. Moreover, there can be no exceptions to the internal relation being shown or displayed. It would make no sense, for example, for me to add, “That is what softness feels like now”. Finally, my statement “That is what softness feels like” is related directly or non-mediatively to the action (performed by the questioner) of squeezing the sponge. There thus can be no third item (either inside my head, inside the head of the questioner or spread out in the environment) that mediates between my statement and the questioner’s action.

These two examples highlight the contextuality of internal relations (a point discussed in Chap. 4, Sect. 4.2). In the first example, there is no internal relation. I am simply squeezing the sponge and experiencing the feeling of softness. In the second example, there is an internal relation. This is because I am defining the experience of softness. Thus, whether or not an internal relation applies in a given context will depend on what is being done in that context.

Continuing with the second example, suppose I wanted to test whether the questioner had grasped my point, that is, whether they could fulfil the behavioural criterion needed for mastery of the concept of softness. I might thus hand them the same sponge sometime later and ask them, "So what does softness feel like?" If the questioner were to squeeze the sponge while saying, "That is what softness feels like", then they would be showing or displaying the same internal relation as before. This would be evidence that they had mastered the relevant concept.

If this is correct, then there is no process actualized when the questioner squeezes the sponge and says, "The sponge feels soft" that makes the questioner's experience one of softness. Rather it is the internal relation that the questioner enacts between their action and their statement that makes their experience one of softness. Internal relations are a temporal and so cannot be understood as processes, since processes are temporal. We should thus reject the Process Model.

Structural Enacted Mind can then avoid the Representationalist Objection. Phenomenal experiences cannot be framed in either causal or constitutive terms. This is because actions are identified with experiences and so neither cause such experiences nor constitute such experiences. In which case, the sort of causal-constitutive distinction that drives RO no longer has application. Moreover, since Structural Enacted Mind rejects the Process Model, and diachronic constitution is linked to that Model, then Structural Enacted Mind also rejects diachronic constitution. Structural Enacted Mind thus provides a way to read Enacted Mind that avoids RO without endorsing diachronic constitution.[6]

However, if Structural Enacted Mind is a version of Enacted Mind, what then is the role of implicit or practical knowledge of sensorimotor dependencies? In Chap. 3 Sect. 3.3, I claimed that Structural Enacted

Mind restricts the role of such knowledge. We can now clarify what this restriction involves.

In Chaps. 4 and 5, I discussed the distinction between the behavioural criteria needed for mastery of a concept (following a rule, new pain behaviour) and how one learns that concept. In the context of the second example mentioned above, the questioner must perceive the internal relation I am showing or displaying via my statement and their action. This is how the questioner comes to learn the phenomenal concept. This perception could be understood in terms of the sorts of practical knowledge favoured by Sensorimotor Enactivists. After all, the internal relation involved here is being shown or displayed and the questioner must perceive this relation.[7] If so, then implicit or practical knowledge would be restricted to the learning of phenomenal concepts (by contrast, no such knowledge would be needed to perceive simpliciter). Nonetheless, as also mentioned above, it is only when the questioner enacts an internal relation themselves that they thereby fulfil the behavioural criterion and so master the concept.

6.4 Structural Extended Mind

In Chap. 2, I outlined two possible readings of Extended Mind. Both readings encountered problems. The problem facing the first reading was specifying the functional roles of biological inner processes and non-biological outer processes. Too precise and there would not be parity between the inner and the outer. Too liberal and there would be cognitive bloat.

I then considered a second reading of Extended Mind, according to which cognitive extension is based on complementarity between what happens inside the head and what happens in the environment. However, this appeal to complementarity failed to demonstrate why coupling between an individual and use of an environmental object ensures that such use is thereby part of and so extends the mind of the individual. This reading could avoid this problem if it was understood as a methodological claim. But then the first reading would have priority over the second.

Following Sutton (2010), I called the first reading First Wave Extended Mind and the second reading Second Wave Extended Mind.

However, Sutton (2010) has also suggested that there could be a Third Wave of Extended Mind. Building on Sutton's suggestion, Kirchhoff (2012) identifies three key themes that might define a Third Wave Extended Mind.

First, a Third Wave rejects the idea that cognitive extension is a sometimes event. Instead the boundaries between the inner and the outer are now understood to be always porous or open-ended. Extension is thus the default position. Second, it rejects the idea that our abilities to represent come from what is inside our heads. Instead these abilities are products of our enculturation within shared social and linguistic practices. Third, it also rejects the idea that cognition is organism centred. Instead cognition is now understood to be emergent from and so distributed among different forms of media.

These three themes then place considerable distance between First and Third Wave Extended Mind.

As I showed in Chap. 2, First Wave is defined by its appeal to the Parity Principle (PP). However, PP only works if we assume that in-the-head processing sets the standard for what counts as cognitive. Yet Clark and Chalmers (1998) provide no argument for this claim. Given the centrality of PP to First Wave, this is a problem. One way for Clark and Chalmers to resolve this problem, so I claimed (see Chap. 2, Sect. 2.2), was to invoke internal representations.

Yet Third Wave both rejects the idea that our abilities to represent come from what is inside our heads and that cognition is organism centred. If so, then, on a Third Wave reading, there can be nothing inside the head that sets the standard for what counts as cognitive. In which case, Third Wave, unlike First Wave, challenges the Representational Orthodoxy.[8]

Kirchhoff and Kiverstein (2018) show what such a "decentralized" view of cognition might look like, which they do so by appealing to, among other things, diachronic constitution. This appeal to diachronic constitution, so they claim, reveals that our enculturation within shared social and linguistic practices is not some background condition needed for cognizing (as would apply on a First Wave reading of Extended Mind).

Rather such practices can play key roles in cognizing. An example of my own will help illustrate how this might work.

I decide to compose a text message to my wife. I have some vague ideas about what I want to say but I am unsure about how I want to say it. Still, I want to get it right, as I want to ensure that my wife fully understands how I feel. So, I begin, "Hi Catherine" and type a few words but then I delete them. I then type a few more, which I then keep. I thus go backwards and forwards in this hesitating manner, typing some text, deleting other text, typing some more, until I have finally composed my message. Once done, I then click send.

Notice that in this example, my composition of the text happens in this hesitating back-and-forth manner. In terms of Third Wave, we could thus regard my cognitive processing as decentralized in the sense that I only loosely coordinate what is happening, since what I want to say is what emerges through my back-and-forth process of typing the message. My cognitive processing is thus a dynamical process, that is, it operates on multiple time-scales, some short term, such as my back-and-forth typing, others long term, such as the eventual message I compose to my wife. Moreover, there is a complex interplay between these short-term and long-term processes, with the short-term processes realizing the long-term processes and the long-term processes feeding back into the short-term processes. We can also see that this dynamical interchange operates within the constraints set by a cultural practice, namely that of sending text messages. Cultural practices thus play more than just a background role. For if there were no cultural practice of sending text messages, then the short and long-term processing evident in this example could not occur.

Notice that this confirms the tacit or implicit link between diachronic constitution and the Process Model. For the above example is what it is because of the processes that occur as and when I type a message to my wife, that is, it is those very processes (both short term and long term) actualized when I type that make what I am doing cognitive. This is to endorse the Process Model.

One might object, however, that given the importance of the cultural practice involved, then it is the practice itself that allows what happens in this example to be regarded as cognitive. But if so, then there is no commitment, tacit or otherwise, to the Process Model. But this objection, if correct, would violate diachronic constitution, given the central importance that diachronic constitution accords to short-term as well as long-term processes. This reiterates a point made earlier (see Sect. 6.2 of this chapter), namely that it is not clear what it would mean to reject the Process Model but still hold on to diachronic constitution.

However, as Chaps. 4 and 5 have demonstrated, Wittgensteinian considerations support rejecting the Process Model. Yet if we reject the Process Model, then this requires that we should also reject Third Wave Extended Mind, given the alleged link between Third Wave and diachronic constitution. On the other hand, as noted in the introduction to this section, both First and Second Wave Extended Mind encounter problems of their own.[9] These difficulties then raise the following question, namely is there an alternative "non-Wave" approach to Extended Mind? I shall argue that there is.

Within the literature on Extended Mind, we can distinguish between, on the one hand, the Hypothesis of Extended Mind (HEM), and on the other, the Hypothesis of Extended Cognition (HEC) (Palermos, 2014; Pöyhönen, 2014; Bernecker, 2014). This distinction is not hard and fast and different philosophers have understood it in various ways. Nonetheless, one way to parse it out is to say that HEC theorists take as given what the mind is and then seek to identify, for explanatory purposes, the boundaries of those sub-personal processes that underpin some given mental episode. In more detail, HEC theorists aim to show that giving an explanation of the sub-personal processes for a particular mental episode requires appealing to an individual's use of an object or objects.

By contrast, HEM theorists focus instead on ontological issues, such as those concerning the boundaries of the mind. Understood in this way, HEM can be read as a version of First Wave Extended Mind. Indeed, as both Palermos (2014) and Pöyhönen (2014) point out, Clark and Chalmers' thought experiment of Otto and Inga is an example of HEM (see Chap. 2, Sect. 2.2), since the thought experiment aims to show how, under conditions of parity or equivalence, the boundaries of Otto's mind could extend outside his head to include Otto's use of his notebook.

Yet if HEM is just another way to read First Wave, and we wish to reject First Wave, then we need to also reject HEM. However, we can still embrace HEC. We can do this if we endorse what I call Structural Extended Mind. According to Structural Extended Mind, experiences and other mental episodes can be identified with actions that involve internal relations. Nonetheless, the sub-personal processes that underpin these actions can be understood as sometimes extending to include our use of objects. I am going to develop this point by focusing on remembering.[10] In particular, I am going to develop a suggestion put forward by Wittgenstein.

For example, Wittgenstein wrote:

> Imagine the following phenomenon. If I want someone to take the note of a text that I recite to him, so that he can repeat it to me later, I have to give him paper and pencil, while I am speaking he makes lines, marks, on the paper; if he has to reproduce the text later he follows those marks with his eyes and recites the text. But I assume that what he has jotted down is not writing, it is not connected by rules with the words of the text; and if anything in it is altered, if part of it is destroyed, he gets stuck in his 'reading' or recites the text uncertainly or carelessly, or cannot find the words at all.—This can be imagined! What I called jottings would not be a rendering of the text, not a translation, so to speak, in another symbolism. The text would not be stored up in the jottings. And why should it be stored up in our nervous system? (RPP I 908)

In this example, there is a correspondence between the lines or marks the individual makes on the page in response to hearing the text and the individual's ability to recite the text, such that any disruption of that correspondence will hinder the individual's ability to recite the text. Nonetheless, it does not follow, insisted Wittgenstein, that the ability to recite the text must be caused by those lines or marks. That is, there need be no continuous chain of events leading from the lines or marks on the page to the individual's ability.[11] Using Structural Extended Mind, I will describe how this might work.

Consider what Wittgenstein says elsewhere:

> We say: "An order orders its own execution." So it knows its execution before it even gets there?—But that was a grammatical proposition, and it

says: if an order runs: "Do such-and-such", then *doing such-and-such* is called "executing the order". (PI 458, emphasis in original)

Using Macha's (2015) account of internal relations (see Chap. 4, Sect. 4.2), we can say that the order, "Do such-and-such" is a grammatical statement, since it shows or displays an internal relation between the statement itself and the action of doing such-and-such. That is, the behavioural criterion for fulfilling the order is enacting that internal relation. I claim that this assessment can be applied to Wittgenstein's example of remembering.

Suppose someone orders the individual to recite the text. This order is grammatical, that is, it shows or displays an internal relation between two complexes, namely an internal relation between the statement of the order to recite the text and the action of reciting the text. The individual then needs to perceive this relation and then show or display the same relation via his or her own behaviour, for example, by reciting the text. The individual has then fulfilled the required behavioural criterion, that is, they have fulfilled the order. This is what it means to remember in this context.

Notice that if this is correct, then there is no process occurring when the individual recites the text that makes what the individual is doing a case of remembering. For if it is the individual's action of showing or displaying an internal relation that makes what they are doing an example of remembering, then their ability to recite the text cannot be understood as a process, since internal relations are a temporal whereas processes are temporal.

However, as also noted in Wittgenstein's example, any disruption to the correspondence between the lines or marks the individual makes on the page will hinder the individual's ability to recite the text. This can be understood in terms of Structural Extended Mind. For the boundaries of the sub-personal processes that underpin the individual's recitation of the text can be understood as extending to include the lines or marks the individual has made on the page. This doesn't involve a commitment to the Process Model. For, as argued above, the processes that extend during those moments that the individual writes down lines or marks on the page are not what make this a case of remembering. Nonetheless, understanding such processes as extending outside the individual's head helps us explain the fact that, as Wittgenstein notes, "if anything in [the lines or marks on the page] is altered, if part of it is destroyed, he gets stuck in his 'reading' or recites the text uncertainly or carelessly, or cannot find the words at all". The individual gets struck because these extending processes are what underpin the individual's ability to recite the text.

This then demonstrates how Structural Extended Mind avoids the Representationalist Objection. For while there is coupling between the individual and the lines and marks on the page, this coupling is not constitutive, since such coupling is not what makes this example one of remembering. In which case, the causal-constitutive distinction that drives RO does not apply. Moreover, given the link between the Process Model and diachronic constitution, Structural Extended Mind's rejection of the Process Model means that it rejects diachronic constitution. Structural Extended Mind thus provides a way to read Extended Mind that avoids RO without endorsing diachronic constitution.

6.5 Full STEEM Ahead

I began this book with the following question: to what extent do 4E approaches actually challenge the Representational Orthodoxy? We now have an answer to this question.

First, I have identified the two conditions that, so I claimed, both Enacted Mind and Extended Mind must fulfil if they are to challenge the Representational Orthodoxy. On the one hand, they must avoid the Representationalist Objection. On the other, they must do so without endorsing diachronic constitution. Second, I then offered a theoretical framework that could meet these two conditions. I called this framework Structural Enacted or Extended Mind or STEEM. STEEM clarifies the anti-representationalist credentials of both Enacted Mind and Extended Mind. As noted in the introduction to this book, 4E then becomes 2E. Yet this is a 2E that arguably fulfils the promise of 4E, namely to deliver a new beginning when it comes to thinking about the mind. Still, we might wonder what this new beginning looks like. In what follows, I want to sketch out one such possibility.

Consider how important smell is. From newly cut grass to freshly brewed coffee, smells are a constant reminder of how embodied and embedded we all are in the world around us. However, what happens in your brain whenever you smell a familiar smell? A standard reply in neuroscience is to say that a group of neurons is activated such that each time you experience that smell, the same group of neurons fires into action. Moreover, so this reply continues, it is the firing of those neurons that explains why you experience the smell as the same and not as a different smell. In the parlance favoured by neuroscientists, this group of neurons represents this familiar smell.

However, new evidence suggests that repeated exposure to the same smell does not activate the same group of neurons. For while a certain group of neurons will be activated during initial exposure to that smell, repeated exposure to the same smell causes that group to change over time resulting in an entirely new group of neurons coming to represent that smell (Schoonover et al., 2021). Neuroscientists have called this phenomenon "representational drift" (ibid.). If this is correct, then the brain's representation of a smell, any smell, can drift over time.

Perhaps this evidence isn't surprising. Given the incredible complexity of the processes within the brain, did anyone really think that a brain's representation of a smell would remain fixed? Yet this response arguably undersells the significance of representational drift. For if the brain's representation of a particular smell can drift over time, then why think that the brain is representing that smell and not another smell? Furthermore, drift is not confined to smell. Place cells in the hippocampus (so called because they fire whenever the individual is located in a particular place) also exhibit drift, that is, the location to which these cells respond also drifts over time (Rubin et al., 2015; Driscoll et al., 2017). Thus, explaining representational drift is a matter of key importance for neuroscience.

As Young (2021) points out, neuroscientists aim to give an account of drift. Some think that representations in the brain still occur but only at a higher level. Drift then is only a low-level feature confined to particular groups of neurons. Others think that drift is evidence of some currently unknown brain mechanism. This mechanism remains unchanging even if particular groups of neurons do not. If either of these accounts is accepted, then drift can be explained. However, as Young continues, such replies suggest that "[n]euroscience's own representations of the brain still have plenty of room to drift" (ibid.).

I think Young is correct. What then is the brain is doing if it is not engaged in building and maintaining representations of the outside world? Notice that this is a question for philosophy of neuroscience rather than for neuroscience per se, given that answering it requires a theoretical framework that allows us to make sense of the role of the brain in human behaviour.

Think back then to the neuroscientific reply discussed above. The fact that the same group of neurons is activated whenever you smell a particular smell is taken to show why that smell is familiar. This reply assumes that the co-variation or correspondence between a group of neurons and a worldly feature demonstrates the cognitive function of that group of neurons. As such, the order of explanation proceeds from the biological inner to the non-biological outer. Groups of neurons track, in a representational sense, worldly features.

However, if we instead endorse STEEM, then that order is reversed, that is, the order of explanation proceeds from the non-biological outer to the biological inner. For STEEM requires that we distinguish between phenomenal experiences and other mental episodes and the sub-personal processes, which include processes in the brain, which underpin those experiences and episodes. The former are what they are due to behavioural criteria. They are thus structural. However, the latter have no criteria and so are non-structural. This distinction ensures that while there can be co-variations or correspondences between experiences/episodes and underlying processes, such co-variations or correspondences do not represent those experiences/episodes. Indeed, we can only make sense of what the brain is doing once we focus our attention on the behavioural criteria that structure our minds, since only then can we determine what underlying brain processes co-vary with or correlate to a particular phenomenal experience and/or mental episode.

STEEM also requires a reappraisal of what we mean when we talk about the brain. This is because STEEM allows that brain processes can sometimes extend to include non-neural elements (see Sect. 6.4 of the present chapter). If so, then the brain cannot be understood as some self-contained organ. For if brain processes can extend beyond the brain, then the boundaries of the skull do not fix the boundaries of the brain. Rather such processes can potentially extend throughout the body[12] and in some circumstances even into the world.

What then is the brain is doing if it is not engaged in building and maintaining representations of the outside the world? If we endorse STEEM, then we can answer this question. The brain is not building and maintaining any representations. Indeed, given that representational drift arises from the idea that there are representations in the brain, then STEEM dissolves the whole issue of representational drift. Instead, if we are to determine the role our brains play in human behaviour, then we first need to examine the many ways behavioural criteria structure our minds. STEEM thus requires that we rethink what is going on inside our heads. In this sense, STEEM can herald a new beginning when it comes to thinking about the mind.

Notes

1. I call this framework Structural Enactive or Extended Mind (STEEM) due to the crucial role that internal relations play within this framework. I take the term structure as it applies to internal relations from Glock (see Glock, 1996, p. 90).
2. This sponge example originally comes from Myin (2003).
3. A related view is Manzotti's (2019) Mind-Object Identity view. According to Manzotti, "one's experience of an object is identical to the object itself" (ibid., p. 1) and "the properties of experience are nothing but physical relative properties instantiated by an object relative to another object (which in our case is the body)" (ibid., p. 7). Mind-Object Identity thus denies that experiences are about anything. As Manzotti puts it, "What is my experience of the apple? It is the apple. *Identity is the new intentionality*" (ibid., p. 8, emphasis in original). This looks akin to Sensorimotor-REC identity. Think of squeezing the sponge. The feeling of softness isn't about squeezing the sponge. Instead, to feel softness just is to squeeze the sponge. However, unlike Mind-Object Identity, Sensorimotor Enactivism and REC don't identify experiences with objects but rather with actions (note that Structural Enacted

Mind also adopts this view). Thus, while there are important parallels between Mind-Object identity and these enactivist positions, these positions nonetheless remain distinct, as Manzotti himself recognizes (see ibid., p. 3).

4. For other parallels between Sensorimotor Enactivism and REC, see Myin and Loughlin (2018).
5. The link between enactivist approaches and Wittgenstein is now well established within the philosophical literature on 4E Cognitive Science. In my own work, I have explored this link (see Loughlin, 2019, 2020b, forthcoming-a, b). For other work also exploring this link, see Heras-Escribano et al., 2015; Hutto, 2013; Hutto et al., 2014; Moyal-Sharrock, 2013; Steiner, 2018. As the text makes clear, Structural Enacted Mind offers a unique contribution to this literature. This is because Structural Enacted Mind uses Wittgensteinian considerations in order to show where identities between action and phenomenal experience come from, namely from the fulfilment of certain behavioural criteria. Also, unlike other enactivist approaches, such as REC, STEEM embraces rather than rejects Extended Mind (see Sect. 6.4 of this chapter).
6. If Structural Enacted Mind is accepted, then this suggests that Wittgensteinian considerations may offer a more fruitful resource for enactivists to use when challenging the Representational Orthodoxy than, say, the sort of existential phenomenology found in Heidegger and Merleau-Ponty.
7. A phylogenetic and ontogenetic story could possibly be told about this know-how. For example, a phylogenetic story could be told about how this know-how potentially grounds our customs or language-games, which then allows an ontogenetic story to be told about how the know-how involved in the perception of internal relations then emerges. Nonetheless, these stories will be distinct from the fulfilment of the behavioural criteria needed for mastery of phenomenal concepts.
8. These considerations might seem to place Third Wave Extended Mind closer to Enacted Mind, given that Enacted Mind also challenges the Representational Orthodoxy. However, an obstacle to this assessment is that Enacted Mind, for example, Sensorimotor Enactivism and REC, regards cognition as organism centred (though of course not organism bound). But if so, then it may not be clear whether Third Wave is an example of either Extended or Enacted Mind. See Loughlin (2020a) where I develop this point.

9. First Wave Extended Mind could also be read as committed to the Process Model. Recall Clark and Chalmers' Parity Principle or PP (see Chap. 2, Sect. 2.2). PP is explicitly about processes.
10. Remembering seems an apt case study to consider here given how central remembering is to Extended Mind, for example, the Otto-Inga thought experiment.
11. Klagge (2014) argues that Wittgenstein distinguishes between a meditative model of causality and a non-mediative model of causality. On a meditative model, "any pair of causally connected events must be connected by a series of events continuous in Space and in Time" (ibid., p. 201, fn33). By contrast, on a non-mediative model, two events can be causally connected without there being such a continuous series of events linking them. If we apply Klagge's analysis to Wittgenstein's example, then Wittgenstein is challenging the idea that a meditative model of causality can be applied to this example. I think this is correct. But if so, then Wittgenstein is doing something radical in this example, namely suggesting that the cause of the individual's ability to recite the text can be internal to that ability. In the terminology developed in this book, it is the individual's action of showing or displaying an internal relation between two complexes that makes what they are doing a case of remembering.
12. Thompson (1989) notes that the human brain exhibits dense and reciprocal connections to the Enteric Nervous System (ENS) (sometimes called the gut brain), which regulates digestion. Given this fact, one could regard the processes that occur in the brain as sometimes extending to include the ENS.

References

Bernecker, S. (2014). How to Understand the Extended Mind. *Philosophical Issues, 24*(1), 1–23. https://doi.org/10.1111/phis.12023

Chalmers, D. (1996). *The Conscious Mind.* Oxford University Press.

Clark, A., & Chalmers, D. (1998). The Extended Mind. *Analysis, 58*(1), 7–19.

Driscoll, L. N., Pettit, N. L., Minderer, M., Chettih, S., & Harvey, C. D. (2017). Dynamic Reorganization of Neuronal Activity Patterns in Parietal Cortex. *Cell, 170*, 986–999.

Glock, H. (1996). *A Wittgenstein Dictionary.* Blackwell Publishers.

Heras-Escribano, M., Noble, J., & de Pinedo, M. (2015). Enactivism, Action and Normativity: A Wittgensteinian Analysis. *Adaptive Behavior*, 1–14. https://doi.org/10.1177/1059712314557364

Hutto, D. (2013). Enactivism, from a Wittgensteinian Point of View. *American Philosophical Quarterly, 50*(3), 281–302.

Hutto, D., & Myin, E. (2013). *Radicalising Enactivism: Basic Minds Without Content*. MIT Press.

Hutto, D., Kirchhoff, M., & Myin, E. (2014). Extensive Enactivism: Why Keep It All in? *Frontiers in Human Neuroscience, 8*, article 706.

Kirchhoff, M. (2012). Extended Cognition and Fixed Properties: Steps to a Third-Wave Version of Extended Cognition. *Phenomenology and the Cognitive Sciences, 11*, 287–308.

Kirchhoff, M., & Kiverstein, J. (2018). *Extended Consciousness and Predictive Processing: A Third Wave View*. Routledge.

Klagge, J. (2014). *Wittgenstein in Exile*. MIT Press.

Levine, J. (1983). Materialism and Qualia: the Explanatory Gap. *Pacific Philosophical Quarterly, 64*, 354–361.

Loughlin, V. (2019). Wittgenstein's Challenge to Enactivism. *Synthese*. https://doi.org/10.1007/s11229-019-02244-3

Loughlin, V. (2020a). Extended Mind. In V. P. Glaveanu (Ed.), *The Palgrave Encyclopedia of the Possible*.

Loughlin, V. (2020b). Why Enactivists Should Care About Wittgenstein. *Philosophia*. https://doi.org/10.1007/s11406-020-00286-3

Loughlin, V. (forthcoming-a). Deep Impact: Wittgenstein's Enduring Enactivist Legacy. In C. Sandis & D. Moyal-Sharrock (Eds.), *Extending Hinge Epistemology*. Anthem Press.

Loughlin, V. (forthcoming-b). Why Wittgenstein Was not a Therapist. *Aufklärung*.

Macha, J. (2015). *Wittgenstein on Internal and External Relations: Tracing All the Connections*. Bloomsbury Academic.

Manzotti, R. (2019). Mind-Object Identity: A Solution to the Hard Problem. *Frontiers in Psychology, 10*, 63. https://doi.org/10.3389/fpsyg.2019.00063

Moyal-Sharrock, D. (2013). Wittgenstein's Razor: The Cutting Edge of Enactivism. *American Philosophical Quarterly, 50*(3), 263–279.

Myin, E. (2003). An Account of Colour Without a Subject? *Behavioural and Brain Sciences, 26*(1), 42–43.

Myin, E., & Loughlin, V. (2018). Sensorimotor Enactive Approaches to Consciousness. In R. J. Gennaro (Ed.), *Routledge Handbook of Consciousness* (pp. 202–215). Routledge.

Nagel, T. (1974). What It Is Like to Be a Bat? *Philosophical Review, 83*, 435–450.

O'Regan, K. (2009). Sensorimotor Approach to Phenomenal Consciousness. In T. Bayne, A. Cleeremans, & P. Wilken (Eds.), *The Oxford Companion to Consciousness* (pp. 588–593). Oxford University Press.

O'Regan, K. (2011). *Why Red Doesn't Sound Like a Bell: Understanding the Feel of Consciousness*. Oxford University Press.

Palermos, S. O. (2014). Loops, Constitution and Cognitive Extension. *Cognitive Systems Research, 27*, 25–41.

Pöyhönen, S. (2014). Explanatory Power of Extended Cognition. *Philosophical Psychology, 27*(5), 735–759.

Rubin, I., Geva, N., Steintuch, L., & Ziv, Y. (2015). Hippocampal Ensemble Dynamics Timestamp Events in Long-Term Memory. *e-life*. https://doi.org/10.7554/eLife.12247.001

Schoonover, C. E., Ohashi, S. N., Axel, R., & Fink, A. J. P. (2021). Representational Drift in Primary Olfactory Cortex. *Nature*. https://doi.org/10.1038/s41586-021-03628-7

Steiner, P. (2018). Reading and Understanding: On Some Differences Between Wittgenstein and 4E Cognitive Science. *Revista Italiana di Filosofia del Linguaggio (RIFL), 13*(2), 124–137.

Sutton, J. (2010). Exograms and Interdisciplinarity: History, the Extended Mind, and the Civilising Process. In R. Menary (Ed.), *The Extended Mind* (pp. 189–226). The MIT Press.

Thompson, W. G. (1989). *Gut Reactions: Understanding Symptoms of the Digestive Tract*. Plenum Press.

Young, E. (2021). Neuroscientists Have Discovered a Phenomenon That They Can't Explain. *Atlantic*. https://www.theatlantic.com/science/archive/2021/06/the-brain-isnt-supposed-to-change-this-much/619145/

References

Adams, F., & Aizawa, K. (2008). *The Bounds of Cognition*. Wiley-Blackwell Publishing Ltd.

Bach-y-Rita, P. (1972). *Brain Mechanisms in Sensory Substitution*. Academic Press.

Baker, G., & Hacker, P. (2009). *Wittgenstein: Rules, Grammar and Necessity: Volume 2 of an Analytical Commentary on the Philosophical Investigations, Essays and Exegesis 185–242* (2nd ed.). Wiley Blackwell.

Ballard, D., Haynoe, M., Pook, P., & Rao, R. (1997). Deictic Codes for the Embodiment of Cognition. *Behavioural and Brain Sciences, 20*, 723–767.

Bernecker, S. (2014). How to Understand the Extended Mind. *Philosophical Issues, 24*(1), 1–23. https://doi.org/10.1111/phis.12023

Block, N. (1978). Troubles with Functionalism. *Minnesota Studies in the Philosophy of Science, 9*, 213–325.

Block, N. (2005). Review of Alva Noë, Action in Perception. *Journal of Philosophy, 102*, 259–272.

Brooks, R. (1991). Intelligence Without Representation. *Artificial Intelligence, 47*, 139–159.

Chalmers, D. (1996). *The Conscious Mind*. Oxford University Press.

Clark, A. (2008a). *Supersizing the Mind: Embodiment, Action and Cognitive Extension*. Oxford University Press.

Clark, A. (2008b). Pressing the Flesh: A Tension in the study of Embodied, Embedded Mind? *Philosophy and Phenomenological Research, 76*(1), 37–58.

V. Loughlin, *4E Cognitive Science and Wittgenstein*, New Directions in Philosophy and Cognitive Science, https://doi.org/10.1007/978-3-030-89463-4

Clark, A. (2009). Spreading the Joy: Why the Machinery of Consciousness Is Still Probably Inside Your Head. *Mind, 118*(472), 963–993.

Clark, A., & Chalmers, D. (1998). The Extended Mind. *Analysis, 58*(1), 7–19.

Crane, T. (2003). *The Mechanical Mind: A Philosophical Introduction to Minds, Machines and Representation* (2nd ed.). Routledge.

Di Paolo, E. (2009). Extended Life. *Topoi, 28*, 9–21.

Drayson, Z., & Clark, A. (2019). Cognitive Disability and Embodied, Extended Minds. In A. Cureton & D. T. Wasserman (Eds.), *The Oxford Handbook of Philosophy and Disability*. https://doi.org/10.1093/oxfordhb/9780190622879.013.10.

Driscoll, L. N., Pettit, N. L., Minderer, M., Chettih, S., & Harvey, C. D. (2017). Dynamic Reorganization of Neuronal Activity Patterns in Parietal Cortex. *Cell, 170*, 986–999.

Frazzetto, G., & Anker, S. (2009). Neuroculture. *Science and Society, 10*, 815–821.

Gallagher, S. (2017). *Enactivist Interventions: Rethinking the Mind*. Oxford University Press.

Glock, H. (1996). *A Wittgenstein Dictionary*. Blackwell Publishers.

Heras-Escribano, M., Noble, J., & de Pinedo, M. (2015). Enactivism, Action and Normativity: A Wittgensteinian Analysis. *Adaptive Behavior*, 1–14. https://doi.org/10.1177/1059712314557364

Holtzman, S., & Leich, C. (Eds.). (2010). *Wittgenstein: To Follow a Rule*. Taylor & Francis Ltd.

Hurley, S. (2010). The Varieties of Externalism. In R. Menary (Ed.), *The Extended Mind* (pp. 101–153).

Hutto, D. (2005). Knowing What? Radical Versus Conservative Enactivism. *Phenomenology and Cognitive Sciences, 4*(4), 389–405.

Hutto, D. (2013). Enactivism, from a Wittgensteinian Point of View. *American Philosophical Quarterly, 50*(3), 281–302.

Hutto, D., & Myin, E. (2013). *Radicalising Enactivism: Basic Minds Without Content*. MIT Press.

Hutto, D., & Myin, E. (2017). *Evolving Enactivism: Basic Minds Meets Content*. MIT Press.

Hutto, D., & Satne, G. (2015). The Natural Origins of Content. *Philosophia, 43*, 521–536.

Hutto, D., Kirchhoff, M., & Myin, E. (2014). Extensive Enactivism: Why Keep It All in? *Frontiers in Human Neuroscience, 8*, article 706.

Kirchhoff, M. (2012). Extended Cognition and Fixed Properties: Steps to a Third-Wave Version of Extended Cognition. *Phenomenology and the Cognitive Sciences, 11*, 287–308.

Kirchhoff, M., & Kiverstein, J. (2018). *Extended Consciousness and Predictive Processing: A Third Wave View*. Routledge.
Klagge, J. (2014). *Wittgenstein in Exile*. MIT Press.
Kripke, S. (1982). *Wittgenstein on Rules and Private Language: An Elementary Exposition*. Harvard University Press.
Levin, J. (2008). Functionalism. In E. N. Zalta (Ed.), *The Stanford Encyclopedia of Philosophy*. Retrieved March 2013, from http://plato.stanford.edu/entries/functionalism
Levine, J. (1983). Materialism and Qualia: the Explanatory Gap. *Pacific Philosophical Quarterly, 64*, 354–361.
Loughlin, V. (2013). Sketch This: Extended Mind and Consciousness Extension. *Phenomenology and the Cognitive Sciences, 12*(1), 41–50.
Loughlin, V. (2014). *Extended Mind, Extended Conscious Mind, Enactivism*. Doctoral thesis for the University of Antwerp.
Loughlin, V. (2019). Wittgenstein's Challenge to Enactivism. *Synthese*. https://doi.org/10.1007/s11229-019-02244-3
Loughlin, V. (2020a). Extended Mind. In V. P. Glaveanu (Ed.), *The Palgrave Encyclopedia of the Possible*.
Loughlin, V. (2020b). Why Enactivists Should Care About Wittgenstein. *Philosophia*. https://doi.org/10.1007/s11406-020-00286-3
Loughlin, V. (forthcoming-a). Deep Impact: Wittgenstein's Enduring Enactivist Legacy. In C. Sandis & D. Moyal-Sharrock (Eds.), *Extending Hinge Epistemology*. Anthem Press.
Loughlin, V. (forthcoming-b). Why Wittgenstein Was not a Therapist. *Aufklärung*.
Loughlin, V., & Zahidi, K. (2017). What Is Left of the Active Externalism Debate? *European Journal of Philosophy*. https://doi.org/10.1111/ejop.12249
Macha, J. (2015). *Wittgenstein on Internal and External Relations: Tracing All the Connections*. Bloomsbury Academic.
Malcolm, N. (1988). *Wittgenstein: Nothing Is Hidden*. Blackwell Publishers.
Manzotti, R. (2019). Mind-Object Identity: A Solution to the Hard Problem. *Frontiers in Psychology, 10*, 63. https://doi.org/10.3389/fpsyg.2019.00063
Marsh, H. (2014). *Do No Harm: Stories of Life, Death and Brain Surgery*. Orion Publishing Group.
Menary, R. (Ed.). (2010a). *The Extended Mind*. The MIT Press.
Menary, R. (2010b). Cognitive Integration and the Extended Mind. In R. Menary (Ed.), *The Extended Mind* (pp. 227–243). The MIT Press.
Miller, A., & Wright, C. (Eds.). (2002). *Rule Following and Meaning*. McGill-Queen's University Press.

Moyal-Sharrock, D. (2013). Wittgenstein's Razor: The Cutting Edge of Enactivism. *American Philosophical Quarterly, 50*(3), 263–279.
Myin, E. (2003). An Account of Colour Without a Subject? *Behavioural and Brain Sciences, 26*(1), 42–43.
Myin, E., & Loughlin, V. (2018). Sensorimotor Enactive Approaches to Consciousness. In R. J. Gennaro (Ed.), *Routledge Handbook of Consciousness* (pp. 202–215). Routledge.
Nagel, T. (1974). What It Is Like to Be a Bat? *Philosophical Review, 83*, 435–450.
Nanay, B. (2018). The Importance of Amodal Perception in Everyday Perception. *i-Perception, 9*(4).
Noë, A. (2004). *Action in Perception*. The MIT Press.
Noë, A. (2012). *Varieties of Presence*. Harvard University Press.
Noë, A., & O'Regan, K. (2002). On the Brain-Basis of Visual Consciousness: A Sensorimotor Account. In A. Noë & E. Thompson (Eds.), *Vision and Mind: Selected Readings in the Philosophy of Perception* (pp. 567–593). A Bradford Book, MIT Press.
O'Regan, K. (2009). Sensorimotor Approach to Phenomenal Consciousness. In T. Bayne, A. Cleeremans, & P. Wilken (Eds.), *The Oxford Companion to Consciousness* (pp. 588–593). Oxford University Press.
O'Regan, K. (2011). *Why Red Doesn't Sound Like a Bell: Understanding the Feel of Consciousness*. Oxford University Press.
O'Regan, K., & Noë, A. (2001). A Sensorimotor Account of Vision and Visual Consciousness. *Behavioural and Brain Sciences, 24*(5), 939–1031.
Palermos, S. O. (2014). Loops, Constitution and Cognitive Extension. *Cognitive Systems Research, 27*, 25–41.
Pöyhönen, S. (2014). Explanatory Power of Extended Cognition. *Philosophical Psychology, 27*(5), 735–759.
Ramsey, W. (2007). *Representation Reconsidered*. Cambridge University Press.
Ramsey, W. (2010). Review of Richard Menary (Ed), The Extended Mind. *Notre Dame Philosophical Reviews*. Retrieved March 2013, from http://ndpr.nd.edu/news/24553/?id=22110
Roberts, T., & Columbetti, G. (2015). Extended Affectivity. *Philosophical Studies, 172*, 1243–1263.
Rowlands, M. (2010). *The New Science of the Mind: From Extended Mind to Embodied Phenomenology*. MIT Press.
Rubin, I., Geva, N., Steintuch, L., & Ziv, Y. (2015). Hippocampal Ensemble Dynamics Timestamp Events in Long-Term Memory. *e-life*. https://doi.org/10.7554/eLife.12247.001

Rupert, R. (2004). Challenges to the Hypothesis of Extended Cognition. *Journal of Philosophy, 101*, 389–428.

Schoonover, C. E., Ohashi, S. N., Axel, R., & Fink, A. J. P. (2021). Representational Drift in Primary Olfactory Cortex. *Nature*. https://doi.org/10.1038/s41586-021-03628-7

Schroeder, S. (2006). *Wittgenstein*. Polity Press.

Seth, A. K. (2014). A Predictive Processing Theory of Sensorimotor Contingencies: Explaining the Puzzle of Perceptual Presence and Its Absence in Synesthesia. *Cognitive Neuroscience, 5*(2), 97–118. https://doi.org/10.1080/17588928.2013.87788

Shapiro, L. (2011). *Embodied Cognition*. Routledge, Taylor and Francis Group.

Sprevak, M. (2009). Extended Cognition and Functionalism. *Journal of Philosophy, 106*(9), 503–527.

Steiner, P. (2018). Reading and Understanding: On Some Differences Between Wittgenstein and 4E Cognitive Science. *Revista Italiana di Filosofia del Linguaggio (RIFL), 13*(2), 124–137.

Sterelny, K. (1990). *The Representational Theory of Mind: An Introduction*. Blackwell.

Sutton, J. (2010). Exograms and Interdisciplinarity: History, the Extended Mind, and the Civilising Process. In R. Menary (Ed.), *The Extended Mind* (pp. 189–226). The MIT Press.

Ter Hark, M. (1990). *Beyond the Inner and Outer: Wittgenstein's Philosophical Psychology*. Kluwer Academic Publishers.

Thompson, W. G. (1989). *Gut Reactions: Understanding Symptoms of the Digestive Tract*. Plenum Press.

Thompson, E. (2018). Review of: Evolving Enactivism: Basic Minds Meets Content. *Notre Dame Philosophical Reviews*.

Tribble, E. (2005). Distributing Cognition in the Globe. *Shakespeare Quarterly, 56*, 135–155.

Vidal, F. (2009). Brainhood, Anthropological Figure of Modernity. *History of the Human Sciences, 22*(1), 5–36.

Wheeler, M. (2005). *Reconstructing the Cognitive World: The Next Step*. MIT Press.

Wheeler, M. (2010). In Defense of Extended Functionalism. In R. Menary (Ed.), *The Extended Mind* (pp. 245–270). The MIT Press.

Wilson, R. (2007). A Puzzle About Material Constitution and How to Solve It: Enriching Constitution Views in Metaphysics. *Philosophers' Imprint, 7*(5).

Wittgenstein, L. (1980). *Remarks on Philosophy of Psychology, Volume 1*. Blackwell Publishers Ltd.

Wittgenstein, L. (1983). *Remarks on the Foundations of Mathematics* (G. H. von Wright, R. Rhess, & G. E. M. Anscombe, Eds.). MIT Press.

Wittgenstein, L. (1992). *Last Writings on the Philosophy of Psychology: The Inner and the Outer* (Vol. 2). Blackwell Publishing.

Wittgenstein, L. (1993). *Philosophical Occasions: 1912–1951* (J. Klagge & A. Nordmann, Eds.). Hackett Publishing Company.

Wittgenstein, L. (2001). *Tractatus Logico-Philosophicus*. Routledge.

Wittgenstein, L. (2009). *Philosophical Investigations* (G. M. Anscombe, P. M. S. Hacker, & J. Schulte, Trans.). Wiley-Blackwell.

Young, E. (2021). Neuroscientists Have Discovered a Phenomenon That They Can't Explain. *Atlantic*. https://www.theatlantic.com/science/archive/2021/06/the-brain-isnt-supposed-to-change-this-much/619145/

Index[1]

[1] Note: Page numbers followed by 'n' refer to notes.

V. Loughlin, *4E Cognitive Science and Wittgenstein*, New Directions in Philosophy and Cognitive Science, https://doi.org/10.1007/978-3-030-89463-4

The manufacturer's authorised representative in the EU is Springer Nature Customer Service Centre GmbH, Europaplatz 3, 69115 Heidelberg, Germany. If you have any concerns regarding our products, please contact ProductSafety@springernature.com

Printed and bound by CPI Group (UK) Ltd, Croydon, CR0 4YY

30/09/2026

02225260-0003